T0257183

Auxiliary Signal Design
for Failure Detection

PRINCETON SERIES IN APPLIED MATHEMATICS

EDITORS

Ingrid Daubechies, *Princeton University*
Weinan E, *Pinceton University*
Jan Karel Lenstra, *Georgia Institute of Technology*
Endre Süli, *University of Oxford*

TITLES IN THE SERIES

*Chaotic Transitions in Deterministic and Stochastic Dynamical Systems:
Applications of Melnikov Processes in Engineering, Physics and Neuroscience*
by Emil Simiu

Selfsimilar Processes by Paul Embrechts and Makoto Maejima

Self-Regularity: A New Paradigm for Primal-Dual Interior Points Algorithms
by Jiming Peng, Cornelis Roos, and Tamás Terlaky

Analytic Theory of Global Bifurcation: An Introduction
by Boris Buffoni and John Toland

Entropy by Andreas Greven, Gerhard Keller, and Gerald Warnecke

Auxiliary Signal Design for Failure Detection
by Stephen L. Campbell and Ramine Nikoukhah

The Princeton Series in Applied Mathematics publishes high quality advanced texts and monographs in all areas of applied mathematics. Books include those of a theoretical and general nature as well as those dealing with the mathematics of specific applications areas and real-world situations.

Auxiliary Signal Design
for Failure Detection

Stephen L. Campbell
and Ramine Nikoukhah

PRINCETON UNIVERSITY PRESS

PRINCETON AND OXFORD

Published by Princeton Univeristy Press, 41 William Street, Princeton, New Jersey, 08540

In the United Kingdom: Princeton University Press, 3 Market Place, Woodstock, Oxfordshire OX20 1SY

The publisher would like to acknowledge the authors of this volume for providing the camera-ready copy from which this book was printed.

Library of Congress Cataloging-in-Publication Data
   Campbell, S.L. (Stephen La Vern)
   Auxiliary signal design for failure detection / Stephen L. Campbell and Ramine Nikoukhah.
      p. cm.
  Includes bibliographical refrences and index.
  ISBN: 0-691-09987-1 (cl: alk.paper)
     1. System failures (Engineering) 2. Fault location (Engineering) 3. Signal processing. I.
  Nikoukhah, Ramine, 1961 – II. Title.
TA169.5.C36 2004
620'.0044–dc21   2003056476

British Library Cataloging-in-Publication Data is available

This book has been composed in Times using LaTeX and the Textures software.

Printed on acid-free paper. ∞

www.pupress.princeton.edu

Printed in the United States of America

10 9 8 7 6 5 4 3 2 1

Contents

Preface

The general areas of failure detection and model identification cover a number of different technical areas and numerous types of applications. No book could hope to cover all of this material. This book focuses on one particular approach to failure detection which is the use of auxiliary signals to assist in multimodel identification.

The approach of this book has a nice geometrical basis. Considerable effort is spent in developing this geometrical insight and providing a logical development of the material. In our development we have paid equal attention to developing a rigorous theory of some generality, a well motivated presentation, and the development of practical numerical algorithms, some of which are given in the last chapter.

The material in this book draws from a number of areas in engineering and applied mathematics, including control theory, numerical analysis, functional analysis, convex analysis, optimization, and the theory of differential equations. The book is potentially of use to a variety of readers from practicing engineers to applied mathematicians. Given the range of potential readers and the number of different areas that are involved, we have had to make a number of decisions. While it should be accessible to others, we have aimed the presentation at the level of a graduate student in *either* engineering or applied mathematics. We have also wanted the material to be readable and not overly pedantic and formal. Thus we have made an attempt to define all terminology that might be known to only a subset of readers. However, these definitions are often included in the text and given in an informal style. We also believe that you do not really understand something unless you have examples in hand, so we have included a large number of specific examples.

We feel strongly that one should have both a theoretical understanding and computational algorithms so the book considers both. The algorithms we develop could be implemented in a number of different environments. For most of the numerical development we have chosen to use the environment Scilab, since it is widely available, free, and exists on a variety of platforms such as Windows, Linux, and most Unix workstations including the newer Macintoshs running OS X. However, one could use any other package that has good linear algebra and numerical integration capabilities, such as MATLAB. Some of the experimental codes we have written are given in chapter 6.

There are different ways to read a paper or a book. One approach is to start at the beginning and carefully develop each fact before going on to the next. A second is to first read the paper to pick up the main results and get a feel for what is going on. Then as need or interest dictates one goes into the proofs or other aspects in more detail. The first approach is the norm in most mathematics papers and books. The second is very common in engineering journals, where proofs are sometimes in the appendixes. After considerable thought, we have chosen to follow a hybrid approach. The chapters proceed in a logical fashion. Within each chapter the presentation is generally logical, starting from the simpler and moving to the more complex. However, in several of the chapters, we make repeated use of some results which are either more technical or use other types of mathematics and whose proof, we felt, would break the flow of the presentation. In these cases, the results needed are stated in the most convenient form for the section they are in, and the general result is stated and proved in a section at the end of the chapter. In this way, the reader trying to first understand the general framework can just follow the presentation and the person interested in all the proofs also has a well laid out path to follow.

One of the hardest decisions we faced concerned referencing. While this book is unique in its approach toward the main topic, the general area of failure detection has a very large number of papers. Since we did not want to have a two hundred page bibliography and still leave some people out, we have gone the other route. In the Introduction, where we give a more general overview of failure detection, we have cited several of the pioneering papers in different areas. The goal in the Introduction is not to give an exhaustive survey but rather to help the reader see how the contributions of this book fit into the overall area and how the area has developed. We have also included references to those papers and books that had an influence on us as we wrote this book. Finally, we have also included a number of references for the reader who is interested in learning more about some of the topics discussed. These references should be viewed only as a starting point. No attempt is made at completeness.

A number of people and institutions have made this book possible. The generous support of the National Science Foundation in the United States and INRIA-Rocquencourt in France has nourished the collaboration of the authors over the last decade. The Boeing Corporation provided the software SOCS used in chapter 4. Colleagues and graduate students have made many contributions ranging from directly collaborating on this research to partaking in discussions while we formulated our ideas. We note here only Dong Kyoung Choe, Francois Delebecque, Kimberly Drake, Kirk Horton, and Bernard Levy.

Finally we would like to thank our wives, Gail Campbell and Homa Nikoukhah, for their support in this and everything else we do.

Steve Campbell
Ramine Nikoukhah

Chapter One

Introduction

1.1 THE BASIC QUESTION

In this book, we study the problem of active failure detection in dynamical systems. Failure detection in some form is now part of essentially every complex device or process. In some applications the detection of failures, such as water losses in nuclear reactors or engine problems on an aircraft, is important for safety purposes. The detection of failure leads to emergency actions by operators or computers and results in a quick and controlled shutdown of the system. In other situations, such as on space missions, the detection of failures results in the use of back-up or alternative systems. These are the more dramatic examples and are often what one first thinks of when hearing of a failure. But in today's society failure detection also plays a fundamental role in managing costs, promoting efficiency, and protecting the environment. It is often much more economical to repair a part during a scheduled maintenance than to have a breakdown in the field. For example, failure may mean that a part or subsystem is not performing to specification, resulting in increased fuel consumption. Detecting this failure means that a scheduled repair can be made with savings of both resources and money. Failure can also mean that a part or subsystem is not performing as expected and that if allowed to continue the result could be a catastrophic failure. But again detection of this type of failure means that repairs can be initiated in an economical and convenient manner. It is much easier to repair a weakened pump than to have to clean up a major sewage spill.

A number of specific examples from applications are in the cited literature. In chapters 2 and 3 we shall use a couple of intuitive examples to motivate some of the ideas that follow. Simpler academic examples will be used to illustrate most of the key ideas and algorithms. Then in the later chapters we shall include some more detailed examples from application areas.

Because of the fundamental role that failure detection plays, it has been the subject of many studies in the past. There have been numerous books [3, 69, 4, 26, 39, 70, 27] and survey articles [83, 44, 1, 38, 34, 68, 35, 36, 37] dedicated to failure detection. The book by Chen and Patton [26] in particular gives an up to date overview of the subject.

Most of these works are concerned with the problem of *passive failure detection*. In the passive approach, for material or security reasons, the detector has no way of acting upon the system. Rather, the detector can only monitor the inputs and the outputs of the system and then try to decide whether a failure has occurred, and if possible of what kind. This decision is made by comparing the measured input-output behavior of the system with the "normal" behavior of the system. The passive approach is often used to continuously monitor the system, although it can also be used to make periodic checks. One simple example of a passive failure detection system is the one that monitors the temperature of your car engine. If the engine gets too hot a warning light may come on. The detector does nothing but passively estimate the engine temperature and compare it to the maximum allowable temperature.

A major drawback with the passive approach is that failures can be masked by the operation of the system. This is true, in particular, for controlled systems. The reason for this is that the purpose of controllers, in general, is to keep the system at some equilibrium point even if the behavior of the system changes. This robustness property, which is clearly desired in control systems, tends to mask abnormal behaviors of the system. This makes the task of failure detection difficult, particularly if it is desired to detect failures that degrade performance. By the time the controller can no longer compensate for the failure, the situation may have become more severe, with much more serious consequences. An example of this effect is the well known fact that it is harder for a driver to detect an underinflated or flat front tire in a car that is equipped with power steering. This trade-off between detection performance and controller robustness has been noted in the literature and has led to the study of the integrated design of controller and detector. See, for example, [60, 80]. A more dramatic example occurred in 1987 when a pilot flying an F-117 Nighthawk, which is the twin-tailed aircraft known as the stealth fighter, encountered bad weather during a training mission. He lost one of his tail assemblies but proceeded back and landed his plane without ever knowing that he was missing part of the tail. The robustness of the control system in this case had the beneficial effect of enabling the pilot to return safely. However, it also had the effect that the pilot did not realize that his aircraft had reduced capability and that the plane would not have performed correctly if a high-speed maneuver was required.

But the problem of masking of failures by the system operation is not limited to controlled systems. Some failures may simply remain hidden under certain operating conditions and show up only under special circumstances. For example, a failure in the brake system of a truck is very difficult to detect as long as the truck is cruising along the road on level ground. It is for this reason that on many roads, just before steep downhill stretches, there are signs asking truck drivers to test their brakes. A driver who disregarded these signs would find out about a brake failure only when he needed to brake going downhill. That is, too late to avoid running off the road or having an accident.

An alternative to passive detection, which could avoid the problem of failures being masked by system operation, is *active detection*. The active approach to failure detection consists in acting upon the system on a periodic basis or at critical times using a test signal in order to detect abnormal behaviors which would otherwise remain undetected during normal operation.

The detector in an active approach can act either by taking over the usual inputs of the system or through a special input channel. An example of using the existing input channels is testing the brakes by stepping on the brake pedal. One class of applications using special channels is when the system involves a collection of pipes or tubes and a fluid or gas is being pumped through the pipes. A substance is injected into the flow in order to determine flow characteristics and pipe geometry. A specific example is the administration of dyes using intravenous injection when conducting certain medical imaging studies. The imaging study lasts for a certain period of time. Since many people react to the dyes, it is desired to keep both the total amount of dye and the rate at which the dye is injected small, consistent with getting sufficient additional resolution.

The active detection problem has been less studied than the passive detection problem. The idea of injecting a signal into the system for identification purposes, that is, to determine the values of various physical parameters, has been widely used and is a fundamental part of engineering design. But the use of extra input signals specifically in the context of failure detection was introduced by Zhang [90] and later developed by Kerestecioğlu and Zarrop [49, 48, 50]. These works served as part of the initial motivation for our study. However, these authors consider the problem in a very different context, which in turn leads to mathematical problems that are very different from those we consider in this book. Accordingly, we have chosen not to review them here.

There are major efforts under way in the aerospace and industrial areas to try to get more extended and more autonomous operation of everything from space vehicles to ships at sea. Regular and extensive maintenance is being replaced by less frequently scheduled maintance and smaller crews. This is to be accomplished by large numbers of sensors and increased software to enable the use of "condition-based maintenance." Active failure detection will play an increasingly important role both in the primary system and in back-up systems to be used in the case of sensor failures.

Before beginning the careful development in chapter 2, we will elaborate a little more on the ideas we have just introduced in this section.

1.2 FAILURE DETECTION

Failure detection consists of deciding whether a system is functioning properly or has failed. This decision process is based on measurements of some of the inputs and outputs of the system. In most cases, these measurements are obtained from sensors placed on or around the system and from knowledge of some of the control inputs.

Given the measurements, the problem is then to decide if the measurement data are consistent with "normal functioning" of the system.

There are two ways of approaching this problem. One is to define a set of input-output trajectories consistent with normal operation of the system. These trajectories are sometimes called the *behavior* of the system. Failure detection then becomes some type of set inclusion test. The other approach consists of assigning a probability to each trajectory and then using probabilistic arguments to build a test. But even in the first approach the notion of probability is often present because without it there is in general no way of defining a set of normal trajectories without being overly conservative. What we often do is to exclude "unlikely" trajectories from this set by selecting an a priori threshold on the likelihood of the trajectories that we admit into the set. Indeed, under the assumption that the observations result from the model, an abnormal behavior is nothing but an unlikely event. There are numerous variations on these two approaches. The choice of which to use is influenced by the nature of the problem.

In *model-based* failure detection, the normal (nonfaulty) behavior of the system is characterized using a mathematical model, which can be deterministic or stochastic. This model then defines an *analytical redundancy* between the inputs and the outputs of the system which can be tested for failure detection purposes. The use of analytical redundancy in the field of failure detection originated with the works of Beard [5] and Jones [46], and of Mehra and Peschon [59] in the stochastic setting. A good picture of the early developments is given in the survey by Willsky [83].

This book develops a model-based approach for several classes of models which consist of differential, difference, and algebraic equations. Accurate models of real physical systems can become quite complex, involving a variety of mathematical objects including partial differential equations (PDEs). However, it is intrinsic to the problem of failure detection that the tests have to be carried out either in real time or close to it. The whole point of failure detection is to determine that a failure has occurred in time to carry out some type of remedial action. Thus, while some calculations, such as design of the detector, can be done off-line, the actual detection test must usually be able to be carried out on-line. To accomplish this, the model used for failure detection purposes in most cases is linear. Nonlinear effects are often included in the noise and model uncertainty effects. In addition, either the models are finite dimensional, or in the case of differential equations, the dimension of the state space is finite. This often requires some type of approximation process if the true underlying models are infinite dimensional. We illustrate this in chapter 4 when we consider differential equation models that include delays.

In the simplest case of a model given by a dynamical system, we would have a deterministic system with a known initial condition. In this case the set of normal behaviors would be reduced to a single trajectory. The failure detection test in this case would be very simple, but this situation does not correspond to real-life cases encountered in practice.

A first step in building more realistic model-based normal behavior sets is to consider that the initial condition of the model, characterizing the behavior set, is unknown. To illustrate, suppose that for a continuous-time differential dynamical system, all the information we have is summarized in the system equations

$$\dot{x} = Ax + Bu, \tag{1.2.1a}$$

$$y = Cx + Du, \tag{1.2.1b}$$

where u and y are, respectively, the measured input and output of the system, x is the state, and A, B, C, D are considered known. In the corresponding discrete-time case, the system equations would be

$$x(k+1) = Ax(k) + Bu(k), \tag{1.2.2a}$$

$$y(k) = Cx(k) + Du(k). \tag{1.2.2b}$$

This way of introducing uncertainty in the model is reasonable because the initial condition of the model usually corresponds to the internal state of the system, which is not directly measured. This approach also leads to simple tests for the inclusion of observed data in the set of normal behaviors. For example, in the discrete-time case, the set of input-outputs satisfying (1.2.2) can be characterized in terms of a set of linear dynamical equations involving only measured quantities u and y. To illustrate, suppose we denote the time shift operator by z. Then the system (1.2.2) can be expressed as follows (using the shift operator is equivalent to taking the z transform of the system, which is the discrete analogue of taking a Laplace transform of a continuous-time system):

$$\begin{pmatrix} -zI + A \\ C \end{pmatrix} x = \begin{pmatrix} -B & 0 \\ -D & -I \end{pmatrix} \begin{pmatrix} u \\ y \end{pmatrix}. \tag{1.2.3}$$

Thus if $H(z)$ is any polynomial matrix in z such that

$$H(z) \begin{pmatrix} -zI + A \\ C \end{pmatrix} = 0, \tag{1.2.4}$$

and the matrix-valued polynomial $G(z)$ is defined by

$$G(z) = H(z) \begin{pmatrix} B & 0 \\ D & -I \end{pmatrix}, \tag{1.2.5}$$

then the relation

$$G(z) \begin{pmatrix} u \\ y \end{pmatrix} = 0 \tag{1.2.6}$$

must hold. The analytical redundancy relations (1.2.6) are also called *parity checks*. They are easy to test at every time step, but they are not unique. In the actual implementation of this approach, the choice of the test is made so as to account for unmodeled model uncertainties, and the result is tested against a threshold. See [53] for one such approach.

The other main method for testing the inclusion of observed data in the set of normal behaviors is to use an *observer*. Observers play a fundamental role in control theory. Given a dynamical system, an observer is a second dynamical system which takes the inputs and outputs of the first system as inputs and whose state (or output) asymptotically approaches the state (part of the state) of the first system. This convergence takes place independently of the initial conditions of the original system and the observer system. If the model is assumed to be perfectly known and only the initial condition is unknown, the observer residual, which is the difference between the measured output and its prediction based on past measurements, converges exponentially to zero. Thus this residual can be used for failure detection testing. Such tests are called observer based. In the continuous-time setting, for example, the observer-based residual generator for system (1.2.1) can be constructed as follows:

$$\dot{\hat{x}} = A\hat{x} + Bu - L(y - C\hat{x}), \quad \hat{x}(0) = 0,$$
$$r = y - C\hat{x} - Du,$$

where r denotes the residual and L is a matrix chosen such that $A + LC$ is **Hurwitz** (all eigenvalues have a negative real part) to assure the convergence of r to zero.

In practice, the residual is tested against a threshold to account for uncertainties. The freedom in the choice of the observer is used for robustness purposes. One such method can be found in [28].

It turns out that the observer-based detection and parity check methods, which historically have been developed independently, are in fact very similar. As is shown in [56], in the discrete-time case, the parity check test is equivalent to an observer-based test where the observer is taken to be deadbeat (L is chosen so that $A + LC$ is nilpotent).

Assuming that all the uncertainties are concentrated in the initial condition alone does not correspond to the reality of most applications. Indeed, the application of the parity check and observer-based methods requires a delicate robustification stage which is constructed using ad hoc methods. In fact, it is the combination of the model and the thresholding test that defines the set of normal behaviors of the system, and not just the dynamic model.

In a model-based approach, rather than focusing on the use of thresholds, it is more natural, and also more informative from a theoretical point of view, to capture the uncertainties using the model itself. A first step in this direction would be to consider additive noise. For example, in the continuous-time setting a model with additive noise might take the form

$$\dot{x} = Ax + Bu + M\nu, \tag{1.2.7a}$$

$$y = Cx + Du + N\nu, \tag{1.2.7b}$$

where ν represents the additive input noise.

In the deterministic setting, the simplest type of noise would correspond to a completely unknown input ν. In this case the set of normal behaviors would be characterized again in terms of a linear model but with both the initial condition and one or more inputs unknown. The test of inclusion for this set of behaviors turns out to be straightforward to build. Both a residual-based method (using unknown input observers or the eigenstructure assignment method developed by Patton and coworkers; see chapter 4 of [26]) and parity check tests (constructed for a reduced model) can be used.

However, the use of unknown additive noise and an unknown initial condition for the model can lead to a very conservative test. That is, it can lead to considering a set of normal behaviors that is much larger than the actual set of normal behaviors. The reason for this is that we usually have some information on the state and on the inputs and outputs of the system. It is rare in a physical system to have a parameter that can take any value a priori. One way to exploit information about variables is to use stochastic modeling and consider the initial condition and the inputs to be random variables. This approach does not directly define a set of normal behaviors. Rather, it associates a probability with each trajectory. The failure detection tests applies a threshold to this probability to decide whether or not a failure has occurred. A set of normal behaviors is implicitly defined as those of high enough probability to exceed the threshold.

Stochastic modeling allows the designer to incorporate realistic information concerning additive noises, in terms not only of the mean and variance, but also of the frequency spectrum. The construction of failure detection tests for such models was introduced in [59], where a complete solution was given based on the Kalman filter. In this approach, the decision is based on statistical tests performed on the Kalman filter's innovation, which in the absence of a fault must be of zero mean, white, and have known variance. In a sense, this method is again observer based because a Kalman filter is just a special case of an observer where the observer gain is tuned to satisfy certain stochastic properties. The innovation here plays the role of the residual in the observer-based method. The difference is that the thresholding test is not ad hoc but corresponds to a statistical property.

Considering a model that contains both unknown inputs and inputs modeled by stochastic processes allows for both robustness and performance, as has been shown in [81], where Kalman filters for the descriptor system were used to deal with unknown inputs. A unified theory of residual generation for models containing unknown inputs and inputs modeled by stochastic processes was developed in [61], where a complete solution was also presented.

Even though it is possible to model many system uncertainties in terms of additive noises, this approach usually leads to conservative designs. More realistic and less conservative models can be constructed by allowing model uncertainty and bounded noise terms in the model. This has been studied in recent years in the context of estimation theory, where the objective is to construct an estimate of the state of the system based on input-output measurements, and is commonly referred to as robust filtering. Robust filtering originated with the works of Bertsekas and Rhodes [8], Schweppe [78, 79], and Kurzhanski and Valyi [52]. It was developed for estimating the states of dynamical models corrupted by unknown but bounded disturbances and noises. In [8] the energy bound is studied for characterization of the uncertainties. For system (1.2.7) on $[0, T]$, for example, the bound would take the form

$$(x(0) - x_0)^T P_0^{-1}(x(0) - x_0) + \int_0^T |\nu|^2 dt < d. \tag{1.2.8}$$

Causal estimation is estimation based on only current and past information. It turns out that the solution to the causal estimation problem for (1.2.7) with the noise bound (1.2.8) is intimately connected to the Kalman filter for this system, where ν is interpreted as a unit-variance zero-mean white noise process. In fact, the estimate and the associated error covariance matrix given by the Kalman filter parameterize an ellipsoid which gives the set of x's consistent with the inequality (1.2.8) in the deterministic problem. The connection between these two problems should not be a total surprise, because in the stochastic setting the left hand side of the inequality (1.2.8) can be interpreted as the negative of the log-likelihood of the noise process.

The study of other types of constraints, for example, instantaneous bounds on the norm of $\nu(t)$, has proved to be more difficult, and only conservative solutions have been proposed. The problem is that pointwise bounds lead to sets that have "corners" and are not even strictly convex. Thus the set of x corresponding to consistent state trajectories is not strictly convex, and the existing analysis requires that bounding ellipsoids must be used. See [78].

In all of the work mentioned so far, it was assumed that the nominal model was perturbed only by additive noises. However, in many real systems the equations themselves are not known exactly or may change with time. There is always some variability in components. In recent years Petersen and Savkin [72, 76], Sayed [77], and El Ghaoui *et al.* [40, 41, 42] have proposed various methods for handling the case where the true model includes both perturbations to the model dynamics and additive unknown signals. They consider models of the form

$$\dot{x} = (A + \delta A)x + (B + \delta B)u + M\nu, \tag{1.2.9a}$$

$$y = (C + \delta C)x + (D + \delta D)u + N\nu, \tag{1.2.9b}$$

where ν represents the additive noise and $(\delta A, \delta B, \delta C, \delta D)$ represent the uncertainties of the system matrices. In real applications different system entries are subject to different amounts of perturbation and some, such as zeros, may not undergo any perturbation at all. Thus it is reasonable to assume that $\delta A = J_1 \Gamma K_1$ where J_1, K_1 provide structure and scaling to the perturbation and Γ is some sort of arbitrary but bounded perturbation. For technical reasons, it turns out to be useful to replace Γ by $\Delta(I - H\Delta)^{-1}$ to give

$$\begin{pmatrix} \delta A & \delta B \\ \delta C & \delta D \end{pmatrix} = \begin{pmatrix} J_1 \\ J_2 \end{pmatrix} \Delta(I - H\Delta)^{-1} \begin{pmatrix} K_1 & K_2 \end{pmatrix},$$

because under certain types of bounding on Δ this leads to a quadratic problem.

In many respects the failure detection problem is close to the estimation problem, and many of the techniques developed for estimation have been used for failure detection. For example, H_∞ filtering has been used in [31, 33, 54], a game theoretic approach in [28], and a set-valued estimation approach by Savkin and Petersen [75] and Petersen and McFarlane [71]. The uncertainty model adopted in this book is closely related to that of [75, 71], which is also sometimes called model validation. This uncertainty model allows for fairly realistic modeling of uncertainties with reasonable conservatism and leads to failure detection tests of acceptable complexity.

Another important method of model-based failure detection uses identification. In the case of parameter identification, the failure detection test is based on the distance between the identified parameter and a nominal parameter corresponding to normal operation of the system. Parameter identification again can be considered as an estimation problem but here the dependence on the parameter is no longer linear and specific estimation methods must be used. We do not use these methods in this book, nor do we use any of the subspace identification techniques recently introduced in the literature; see, for example, [2].

Finally, there have been attempts at using nonlinear models for the purpose of failure detection. Up until now, there have not been any systematic design procedures proposed which could be used efficiently in practice for all nonlinear systems. See the survey paper [35] by Frank. In the special case of bilinear systems, the study can be carried somewhat further; see, for example, [51, 87, 85, 89, 88]. In this book, we initially avoid using nonlinear models. When we have nonlinearities in the system, we try to model them as uncertainties. This approach can be quite conservative when we are dealing with large nonlinearities but it allows us to use powerful tools from linear algebra in applying our methodology. In chapter 4, we introduce an optimization-based method which does allow us to consider some special classes of nonlinear systems.

Before continuing, it is important to comment on our use of the word *conservative*. In much of the failure detection literature an approach being conservative means that either it will avoid making false detection of a failure at the risk of missing some failures or it will avoid missing failures at the risk of false detection of failure (a false alarm). When the active approach presented here is appropriate, that is, the noise bounds and other assumptions hold, the approach provides guaranteed detection. There are no false alarms and no failures are missed. Here the conservatism arises in the size of the auxiliary test signal. When the set of allowable disturbances is increased on applying the theory and algorithms, the computed auxiliary signal may be larger than is required by the original set of disturbances.

1.3 FAILURE IDENTIFICATION

The failure identification problem goes a step beyond failure detection. It concerns not only deciding whether or not a failure has occurred but also, if a failure occurs, determining what kind of failure has occurred. This requires modeling the behaviors of the system input-output trajectories for every possible failure, in addition to the behavior associated with the normal (nonfaulty) system.

In the model-based approach, very often failed systems are modeled in the same way as the normal system but with additional additive inputs representing the effects of various faults. These inputs are assumed either arbitrary and a priori unknown, or constant with known or unknown values, or stochastic. This type of model was already considered in the work of Beard [5].

Even in the case of a single failure, failure detection can be improved by taking into account the model of the failure. For example, by assuming that a failure corresponds to an additive constant input, Willsky and Jones [84] have developed an interesting detection test based on the generalized likelihood ratio or GLR method.

Even when different failures are modeled as unknown inputs, it may be possible under some conditions to identify which failure has occurred if they enter the system in different directions. A nice geometrical theory has been developed by Massoumnia [57]. See also [58]. This characterization is useful in both the deterministic and the stochastic settings. See also [61].

In practice, except for certain types of incipient faults, which are those that appear gradually, the failure cannot be modeled efficiently as an additive input. Consider for example a sensor failure in a system. In the absence of failure the output measurement might be given by

$$y = cx_i + \nu, \tag{1.3.1}$$

where ν represents the measurement noise, x_i is a component of the state measured by the sensor, c is the sensor gain, which is assumed known, and y is the output of the sensor. When the sensor fails, the sensor gain becomes zero. Thus (1.3.1) becomes

$$y = \nu. \tag{1.3.2}$$

The way unknown input failure modeling could be used in this context would be to have

$$y = cx_i + \nu + f \tag{1.3.3}$$

10 CHAPTER 1

represent the behavior of the failed system, where f is assumed to be totally arbitrary. This works out because we can take

$$f = -cx_i, \tag{1.3.4}$$

which makes (1.3.3) equivalent to (1.3.2). Note, however, that this is a conservative approach, because if we have any information about ν (for example, boundedness or statistical properties) (1.3.2) and (1.3.3) are not equivalent. In (1.3.2) we have some information about y, whereas in (1.3.3) y is totally arbitrary.

Another approach to failure identification is to consider a separate model associated with each failure. The models need not even have the same state dimensions. This multimodel approach allows realistic modeling of failures in many applications. For example, it would perfectly capture the situation in the sensor failure problem discussed above. The multimodel approach has been used in particular in the GLR context by Willsky; see, for example, chapter 2 of [3].

In this book, we use the multimodel approach for modeling system failures.

1.4 ACTIVE APPROACH VERSUS PASSIVE APPROACH

There are basically two approaches to failure detection and isolation. One is the passive approach, where the detector monitors input-outputs of the system and decides whether a failure has occurred, and if possible of what kind. A passive approach is used for continuous monitoring and when the detector has no way of acting upon the system. The other approach is the active approach, where the detector acts upon the system on a periodic basis or at critical times, using a test signal called the *auxiliary signal*, over a test period, in order to exhibit possible abnormal behaviors. The decision on whether or not the system has failed, and the determination of the type of failure, are made at the end of this period. Sometimes the tests permit early detection, that is, the decisions are made before the end of the test period. The major theme of this book is the design of auxiliary signals for failure detection.

The structure of the active failure detection method considered here is illustrated in figure 1.4.1. The auxiliary signal v injected into the system to facilitate detection is part (or all) of the system input used by the detector for the period of testing. The signal u denotes the remaining inputs measured on-line, just as the outputs y are measured on-line. In some applications the time trajectory of u may be known in advance, but in general the information regarding u is obtained through sensor data in the same way that it is done for the output y.

In order to simplify this introductory discussion, suppose that there is only one possible type of failure. Then in the multimodel approach we have two sets of input-output behaviors to consider and hence two models. The set $\mathcal{A}_0(v)$ is the set of input-outputs $\{u, y\}$ associated with model 0 of normal behavior. The set $\mathcal{A}_1(v)$ is the set of input-outputs associated with model 1 of behavior when failure occurs. These sets represent possible/likely input-output trajectories for each model. Note that while model 0 and model 1 can differ greatly in size and complexity, the variables u and y have the same dimension in both models.

The problem of auxiliary signal design for guaranteed failure detection is to find a

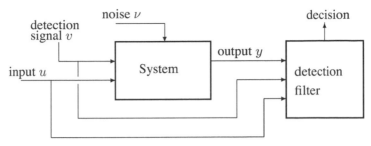

Figure 1.4.1 Active failure detection.

reasonable v such that

$$\mathcal{A}_0(v) \cap \mathcal{A}_1(v) = \emptyset.$$

That is, any observed pair $\{u, y\}$ must come from only one of the two models. Here "reasonable v" means a v that does not perturb the normal operation of the system too much during the test period. This means, in general, a v of small energy applied over a short test period. However, depending on the application, "reasonable" can imply more complicated criteria.

Figures 1.4.2 through 1.4.4 should give a clear picture of the situation. When v is zero, the two sets $\mathcal{A}_0(v)$ and $\mathcal{A}_1(v)$ usually overlap. In particular, when the two sets are associated with linear models, they both contain the origin, as illustrated in figure 1.4.2.

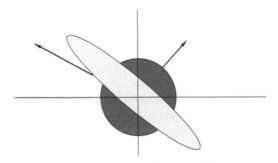

Figure 1.4.2 Auxiliary signal equals zero.

When a signal v other than zero is used, the two sets are moved somewhat apart as illustrated in figure 1.4.3. Increasing the size of v moves the sets further, and at some point they become disjoint as in figure 1.4.4. At this point, we have an auxiliary signal which can be used for guaranteed failure detection. We call such an auxiliary signal *proper*.

The main objective of this book is to present a methodology for the construction of *optimal* proper auxiliary signals in the multimodel context. The auxiliary signal design problem in the multimodel setting has been studied in the past, in particular by Zhang [90] and by Kerestecioğlu and Zarrop [49, 50]. See also the book [48].

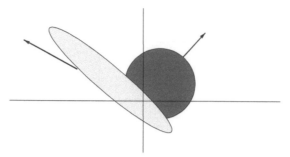

Figure 1.4.3 Small auxiliary signal.

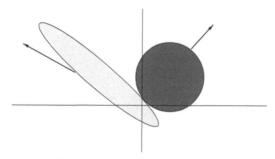

Figure 1.4.4 Proper auxiliary signal.

Our approach, however, differs in many ways from theirs, which was directly motivated by the work on auxiliary signal design for identification purposes and aimed at constructing stationary stochastic auxiliary signals.

We do not restrict our attention to the stationary case; *au contraire*, we focus on short detection intervals where the effects of initial conditions cannot be ignored. In our scenario, the test is to be applied at specific times on a short time interval, perturbing system operation as little as possible, and guaranteeing that if the system has failed, at the end of the test period, the detector discovers the failure.

Our approach is fundamentally deterministic and uses the set membership approach (even if the sets can be based on likelihood thresholding) and we seek guaranteed detectability, a concept that was first introduced in [62]. The work presented in this book follows some of the ideas presented in [62]. However, the models used here are different, thus yielding a very different mathematical theory.

We consider models with different types of uncertainties and in particular realistic model uncertainties, which are often encountered in practice. We have a system theoretic approach and use well established tools such as Riccati equations that allow us to handle very large multivariable systems. The methodology we develop for the construction of the optimal auxiliary signal and its associated test can be implemented easily in computational environments such as Scilab [16, 25] and MATLAB. Moreover, the on-line detection test that we obtain is similar to some existing tests based on Kalman filters and is easy to implement in real time.

This book represents the culmination of several years of research. During this time the results have naturally evolved and have become more general and more powerful. Some of this development and some of the material in this book has appeared in the papers [62, 66, 21, 22, 20, 67, 63, 17, 18, 65, 64, 24].

1.5 OUTLINE OF THE BOOK

In chapter 2, we present the type of models we consider and, in particular, the way that uncertainty can be accounted for in our approach. We also present on-line detection tests associated with these models and discuss their implementation.

The main problem we consider in this book is the construction of the auxiliary signal and its use in failure detection. This problem is considered in chapter 3, where we develop a complete theory and discuss implementation issues. Chapter 3 provides a careful development of the two-model case. The approach of chapter 3 can be applied to more than two models by performing a sequence of two-model tests. When there are more than two possibilities, care must be taken in interpreting each of the sequential tests. This is discussed in chapter 4.

Chapter 4 presents a different approach to the problem of auxiliary signal design. This method is based on numerical optimization. It is less efficient in dealing with large systems, which can be handled by the method presented in chapter 3, but it allows for the consideration of more general models and more complicated constraints. In chapter 3, only linear finite-dimensional models subject to a particular class of uncertainties are considered. In chapter 4, we allow for certain nonlinear models, delays, and more general types of uncertainties. In addition, in chapter 4, we consider an arbitrary finite number of failure models and show how to construct minimal energy proper signals to test for several different failures at once. These simultaneous tests are much more efficient than separate sequential tests. The approach of chapter 4 can also be used for a variety of constrained problems. For example, it is shown how to construct the auxiliary signal of smallest norm for the important case where the auxiliary signal also satisfies a pointwise bound.

Chapter 5 briefly discusses some of the open questions and problem areas for auxiliary signal design.

In chapter 6 we give a collection of programs written in the Scilab language that carry out the algorithms of chapter 3. As described in chapter 6, these are not to be considered as polished industrial grade software. They have been tested on a number of examples.

Chapter Two

Failure Detection

2.1 INTRODUCTION

All approaches to failure detection require some assumption about the underlying process. We consider model-based failure detection. This means we assume that a mathematical model exists describing the normal (unfailed) operation of the system. It is also assumed that any behavior of the model inconsistent with this model implies the occurrence of a failure. In practice, this determination always takes place in the presence of uncertainty, such as model and measurement error, and various disturbances such as noise. Some assumptions must also be made on this uncertainty. Later in this book we shall consider a number of related issues such as having models for particular failures and also identifying the type of failure if a failure has occurred. This chapter introduces the type of model used in the methodology developed in later chapters and gives a solution to the on-line detection problem.

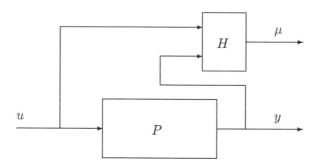

Figure 2.1.1 General setup for failure detection.

The basic problem is as follows. Suppose that we have a plant or process P for which we know the inputs u and outputs y. We wish to decide, based on $\{u, y\}$, whether the plant has failed or not. That is, we want a function $\mu = H(u, y)$ so that the value of μ, often called a *residual*, tells us whether or not the observed $\{u, y\}$ could have been produced by a normal (not failed) P. We call such a $\{u, y\}$ *realizable*.

Often we shall use signal-flow diagrams to illustrate the problem under consideration. The diagram in figure 2.1.1 illustrates the failure detection procedure. Signal-flow diagrams are common in engineering and computer science but are much less frequently used in applied mathematics. The diagrams are easy to interpret. The direction of the lines indicates whether a variable is an input or an output. If a line splits, then the same value goes down both lines. Thus figure 2.1.1 says that $\mu = H(u, y)$ and $y = P(u)$.

We will consider a number of different uncertainty formulations that will use finite-dimensional linear systems involving both differential and difference operators. However, it is easy to lose track of the basic idea in the analysis needed in these more complicated problems. In order to get across the ideas in the simplest possible way, we start by considering the static case. This is the case when the processes have no dynamics and are simply functions between vector spaces. Thus P merely maps a vector to another vector.

In later sections we shall consider the general problem of time-varying linear systems over an interval $[0, T]$, in both continuous and discrete time, and discuss the infinite-horizon case $(T = \infty)$ for time-invariant systems.

2.2 STATIC CASE

We assume in this section that the input u and output y are simply vectors in some finite dimensional vector spaces which we take to be real. All the basic matrices describing the system are also taken to be real.

2.2.1 No Uncertainty

Even though it is not realistic, it is instructive to consider the situation where the behavior of the linear system P is completely known. That is, there is no model uncertainty or noise present. This means that the system behavior is described by

$$y = Gu. \tag{2.2.1}$$

This equation completely defines the set of $\{u, y\}$ corresponding to normal operation of the system. Thus realizability of $\{u, y\}$ is equivalent to $y - Gu = 0$.

Let

$$\mu = T(y - Gu) = H \begin{pmatrix} u \\ y \end{pmatrix}, \tag{2.2.2}$$

where

$$H = T \begin{pmatrix} -G & I \end{pmatrix} \tag{2.2.3}$$

and T is any invertible matrix. Clearly, realizability of $\{u, y\}$ is equivalent to μ being zero. So for this example the failure detection test consists in verifying whether or not $\mu = 0$.

Of course, all real systems have uncertainty. This uncertainty can enter in a number of ways. It can consist of noise in measuring u or y. It can consistent of perturbations in the coefficients of G. There can be unmodeled effects. The type of uncertainty present has a great impact on the form of the detection tests. We will now illustrate several of the more important types of uncertainty on the static problem.

2.2.2 Additive Uncertainty

One of the more common noise models is additive uncertainty. Additive uncertainty may be thought of as a noise or disturbance added to either the input u or the output y or both. In all these cases, P can be modeled as

$$y = G_1 u + G_2 \nu, \tag{2.2.4}$$

where G_1 and G_2 are matrices, and ν is the noise. This situation is diagrammed in figure 2.2.1.

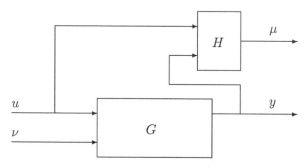

Figure 2.2.1 Additive uncertainty.

Some assumptions will usually have to be made on the noise ν in order to proceed. We shall suppose that there is a bound on the 2-norm of ν. That is,

$$|\nu|^2 < d. \tag{2.2.5}$$

Then $\{u, y\}$ will be realizable if there exists a ν satisfying (2.2.5) and (2.2.4). What is needed is a function just in terms of $\{u, y\}$ which determines realizability.

We assume that G_2 has full row rank. If this were not the case, we would have some non-noisy relations between u and y, which can be treated separately. We shall discuss this possibility later. The full rank assumption on G_2 implies, in particular, that for any $\{u, y\}$ there exists at least one ν satisfying (2.2.4). To find out if among all such ν's there is one that also satisfies (2.2.5), we consider the optimization problem

$$\gamma(u, y) = \min_{\nu} |\nu|^2 \tag{2.2.6a}$$

subject to

$$y = G_1 u + G_2 \nu. \tag{2.2.6b}$$

The function $\gamma(u, y)$ provides the realizability test. If $\gamma(u, y) < d$, then $\{u, y\}$ is realizable. If $\gamma(u, y) \geq d$, then $\{u, y\}$ is not realizable. In order to be useful, we must express γ directly in terms of $\{u, y\}$.

Lemma 2.2.1 *The solution to the optimization problem (2.2.6) is given by*

$$\gamma(u, y) = \begin{pmatrix} u \\ y \end{pmatrix}^T \begin{pmatrix} -G_1^T \\ I \end{pmatrix} (G_2 G_2^T)^{-1} \begin{pmatrix} -G_1 & I \end{pmatrix} \begin{pmatrix} u \\ y \end{pmatrix}. \tag{2.2.7}$$

Proof. Rewrite (2.2.6b) as

$$\begin{pmatrix} -G_1 & I \end{pmatrix} \begin{pmatrix} u \\ y \end{pmatrix} = G_2 \nu. \tag{2.2.8}$$

The optimization problem (2.2.6a), (2.2.8) is a constrained least squares problem. Since G_2 has full row rank, all solutions of (2.2.8) are given by

$$\nu = G_2^\dagger \begin{pmatrix} -G_1 & I \end{pmatrix} \begin{pmatrix} u \\ y \end{pmatrix} + (I - G_2^\dagger G_2)z, \tag{2.2.9}$$

where $G_2^\dagger = G_2^T (G_2 G_2^T)^{-1}$ is the Moore-Penrose inverse of G_2 [23] and z is arbitrary. Since the two summands on the right in (2.2.9) are orthogonal, the minimum norm occurs when $(I - G_2^\dagger G_2)z = 0$. Formula (2.2.7) follows from substituting (2.2.9) with $z = 0$ into $|\nu|^2 = \nu^T \nu$ and using $(G_2^T (G_2 G_2^T)^{-1})^T G_2^T (G_2 G_2^T)^{-1} = (G_2 G_2^T)^{-1}$. $\qquad\square$

Thus if we let

$$\mu = H \begin{pmatrix} u \\ y \end{pmatrix} \tag{2.2.10}$$

where H satisfies

$$H^T H = \begin{pmatrix} -G_1^T \\ I \end{pmatrix} (G_2 G_2^T)^{-1} \begin{pmatrix} -G_1 & I \end{pmatrix}, \tag{2.2.11}$$

we have $\gamma(u, y) = |\mu|^2$. Then the realizability test becomes

$$|\mu|^2 < d. \tag{2.2.12}$$

If the test (2.2.12) fails, then we have a failure.

Example 2.2.1 *Consider the following simple example:*

$$y = u + \nu, \quad \nu^2 < 1. \tag{2.2.13}$$

The realizable set $\{u, y\}$ is given by

$$\gamma(u, y) = (y - u)^2 < 1 \tag{2.2.14}$$

and is illustrated in figure 2.2.2.

Before turning to the discussion of model uncertainty, we wish to make three comments about additive uncertainty.

Stochastic Interpretation

In this book we are interested in deterministic models. However, it is interesting to note that sometimes our results have a stochastic interpretation. Let us modify our detection problem by assuming that ν is a unit-variance zero-mean random vector instead of a norm-bounded unknown vector. In this case, it is no longer possible to construct a feasibility test as we have done before because ν can take any value. Only statistical tests can be used to decide on the probability of failure; that is, the probability that the observed (u, y) has not been generated by our model. It turns out that in this case the μ constructed in (2.2.10) has nice properties.

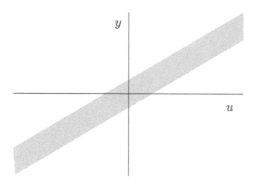

Figure 2.2.2 The realizable set $\{u, y\}$ for example 2.2.1.

Lemma 2.2.2 *Suppose ν in (2.2.4) is a unit-variance zero-mean random variable and μ is defined by (2.2.10) where H satisfies (2.2.11) and has full row rank. Then μ also has zero mean and unit variance.*

Proof. Clearly,

$$\ker(H) = \ker \begin{pmatrix} -G_1 & I \end{pmatrix} = \operatorname{Im} \begin{pmatrix} I \\ G_1 \end{pmatrix}, \tag{2.2.15}$$

where ker and Im denote the kernel (nullspace) and image (range), respectively. Let $\mathcal{E}$ denote the expected value. Then

$$\begin{aligned} \mathcal{E}(\mu) &= \mathcal{E}\left(H \begin{pmatrix} u \\ y \end{pmatrix} \right) = \mathcal{E}\left(H \begin{pmatrix} I & 0 \\ G_1 & G_2 \end{pmatrix} \begin{pmatrix} u \\ \nu \end{pmatrix} \right) \\ &= \mathcal{E}\left(H \begin{pmatrix} 0 \\ G_2 \end{pmatrix} \nu \right) = 0. \end{aligned} \tag{2.2.16}$$

Thus μ has zero mean.

The covariance matrix M of μ is

$$\begin{aligned} M = \mathcal{E}(\mu\mu^T) &= \mathcal{E}\left(H \begin{pmatrix} 0 \\ G_2 \end{pmatrix} \nu\nu^T \begin{pmatrix} 0 \\ G_2 \end{pmatrix}^T H^T \right) \\ &= H \begin{pmatrix} 0 & 0 \\ 0 & G_2 G_2^T \end{pmatrix} H^T. \end{aligned} \tag{2.2.17}$$

Thus

$$\begin{aligned} MH &= H \begin{pmatrix} 0 & 0 \\ 0 & G_2 G_2^T \end{pmatrix} \begin{pmatrix} -G_1^T \\ I \end{pmatrix} \left(G_2 G_2^T \right)^{-1} \begin{pmatrix} -G_1 & I \end{pmatrix} \\ &= H \begin{pmatrix} 0 \\ I \end{pmatrix} \begin{pmatrix} -G_1 & I \end{pmatrix} = H \begin{pmatrix} 0 & 0 \\ -G_1 & I \end{pmatrix}. \end{aligned} \tag{2.2.18}$$

Then using the fact that

$$H \begin{pmatrix} I \\ G_1 \end{pmatrix} = 0 \tag{2.2.19}$$

we get that

$$MH = H \begin{pmatrix} 0 & 0 \\ -G_1 & I \end{pmatrix} + H \begin{pmatrix} I & 0 \\ G_1 & 0 \end{pmatrix} = H. \qquad (2.2.20)$$

But H has full row rank. Thus $M = I$. $\qquad\qquad\qquad\qquad\qquad\qquad$ □

Thus if no failure has occurred ($\{u, y\}$ is generated by our model), the residual μ constructed based on deterministic assumptions has unit variance and zero mean. Thus the statistical test for failure detection would consist in deciding whether or not μ has unit variance and zero mean. What is interesting about this test is that it is generic and completely system independent. This approach allows us to completely decouple the failure detection procedure into two phases. The first phase is generation of μ. The second phase is a generic, system-independent, statistical test.

G_2 not of Full Row Rank

To simplify the discussion earlier we assumed that G_2 in (2.2.4) had full row rank. Suppose that this is not the case. Then G_2 not having full row rank implies that the system exhibits relations between u and y without any noise. If G_2 is not of full row rank, then we can find an invertible matrix T such that

$$TG_2 = \begin{pmatrix} T_1 \\ T_2 \end{pmatrix} G_2 = \begin{pmatrix} G_{21} \\ 0 \end{pmatrix}, \qquad (2.2.21)$$

where G_{21} has full row rank. This is called *row compression* of G_2. In fact, one can take T to be an orthogonal matrix. If we now premultiply (2.2.4) by T, we obtain

$$T_1 y = G_{11} u + G_{21} \nu, \qquad (2.2.22)$$
$$T_2 y = G_{12} u, \qquad (2.2.23)$$

where $G_{1i} = T_i G_1$, $i = 1, 2$. The realizability of $\{u, y\}$ can now be tested using both (2.2.22) and (2.2.23) by generating corresponding residuals μ_1 and μ_2. The second case corresponds to a problem with no uncertainty and the first to a problem with additive, norm-bounded, uncertainty. Realizability of $\{u, y\}$ is then equivalent to $|\mu_1|^2 \leq d$ and $\mu_2 = 0$ both holding.

Note that this problem resulted in our having two different residuals μ_1 and μ_2. This opens the possibility of trying to use the information about which residuals are satisfied and which are not to get additional information about the nature of the failure. We shall return to this point in later chapters.

Unbounded Additive Noise

In some cases, we have a situation where there is no a priori bound available on the noise, or perhaps on a part of the noise. Consider, for example,

$$y = G_1 u + G_{21} \nu_1 + G_{22} \nu_2 \qquad (2.2.24)$$

with

$$|\nu_1|^2 \leq 1 \qquad (2.2.25)$$

and suppose there is no constraint on ν_2. Since ν_2 is completely arbitrary, no information can be obtained about any vectors in the range of G_{22}. If G_{22} has full row rank, then y is arbitrary and any y is realizable. No failure detection is possible. So assume that G_{22} does not have full row rank. Thus we must eliminate, or project out, the range of G_{22}. Let $G_{22}^{\perp}$ be a full row rank matrix of maximal rank such that $G_{22}^{\perp}G_{22} = 0$. $G_{22}^{\perp}$ is sometimes referred to as a maximal rank left annihilator of G_{22}. We can eliminate ν_2 from (2.2.24) by premultiplying (2.2.24) by $G_{22}^{\perp}$. This does not affect the constraints on (ν_1, u, y) implied by (2.2.24). We thus obtain

$$G_{22}^{\perp}y = G_{22}^{\perp}G_1 u + G_{22}^{\perp}G_{21}\nu_1. \tag{2.2.26}$$

If $G_{22}^{\perp}G_{21}$ has full row rank, then lemma 2.2.1 applies, except that G_2 is replaced by $G_{22}^{\perp}G_{21}$ and $\begin{pmatrix} -G_1 & I \end{pmatrix}$ is replaced by $\begin{pmatrix} -G_{22}^{\perp}G_1 & G_{22}^{\perp} \end{pmatrix}$. If $G_{22}^{\perp}G_{21}$ is not of full row rank, then we may proceed as in (2.2.21).

2.2.3 Model Uncertainty

In practice, not only is there noise in outputs and inputs but there is often uncertainty in the models themselves. Developing techniques to deal with some types of model uncertainty is a major challenge in the area of robust control. For a given static plant not all parts of the plant have the same amount of uncertainty and some parts may have no uncertainty at all. This means that P can be modeled as follows:

$$y = (G + \tilde{G})u, \tag{2.2.27}$$

where $\tilde{G}$ represents a structured uncertainty matrix. For example, an entry of $\tilde{G}$ may always be zero since it represents some sort of structural property.

At first glance it would seem natural to assume $\tilde{G} = G_{12}\Gamma G_{21}$, where Γ is an unknown matrix that satisfies some type of bound. However, in developing the theory and the algorithms, it turns out to be better to parameterize the uncertainty in a more general manner which does not lead to any additional complexity. We shall consider the case where $\tilde{G}$ can be modeled as

$$\tilde{G} = G_{12}\Delta(I - G_{22}\Delta)^{-1}G_{21}, \tag{2.2.28}$$

where Δ is any matrix such that

$$\bar{\sigma}(\Delta) \leq 1. \tag{2.2.29}$$

Here $\bar{\sigma}(\Delta)$ denotes the largest singular value of Δ. In (2.2.29) bounds other than 1 are easily rewritten into the bound-1 case by rescaling of the coefficient matrices G_{ij}.

By placing the expression (2.2.28) for $\tilde{G}$ in (2.2.27), we see that the uncertainty in the static plant takes the form of P being modeled by

$$y = (G_{11} + G_{12}\Delta(I - G_{22}\Delta)^{-1}G_{21})u. \tag{2.2.30}$$

For (2.2.30) to be properly defined, we need an additional condition to hold, such as

$$\bar{\sigma}(G_{22}) < 1, \tag{2.2.31}$$

which we assume for the remainder of this section. Condition (2.2.31) guarantees that the inverse exists.

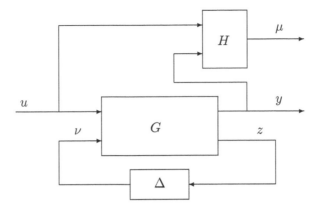

Figure 2.2.3 Model uncertainty.

This way of representing uncertainty is common in the robust control literature. By introducing additional vectors ν and z, we can represent P as is illustrated in figure 2.2.3. In terms of equations we have

$$y = G_{11}u + G_{12}\nu, \tag{2.2.32a}$$

$$z = G_{21}u + G_{22}\nu \tag{2.2.32b}$$

with

$$\nu = \Delta z. \tag{2.2.32c}$$

If (2.2.29) and (2.2.31) hold, then $I - G_{22}\Delta$ is invertible. Then substituting (2.2.32c) into (2.2.32b) and solving for z gives

$$z = (I - G_{22}\Delta)^{-1}G_{21}u. \tag{2.2.33}$$

Then using (2.2.32c) and (2.2.33) for ν in (2.2.32a) gives (2.2.30). Thus (2.2.30) is the form that the uncertainty takes if we view the uncertainty as another system coupled to our original system.

We again wish to develop a test for realizability that depends only on $\{u, y\}$. As in the previous examples, this will be done by setting up an optimization problem. Since $\bar{\sigma}(\Delta) \leq 1$ we have that (2.2.32c) implies that $|\nu|^2 \leq |z|^2$. Hence (2.2.32c) may be replaced by

$$|\nu|^2 - |z|^2 \leq 0. \tag{2.2.34}$$

The system (2.2.32a), (2.2.32b), (2.2.34) is the one we shall work with. Here u, y are assumed known, and ν and z are called, respectively, the *uncertainty input* and *uncertainty output*.

Given $\{u, y\}$, we wish to have a test to see if it could have been produced by the model. To do this, we consider the optimization problem

$$\gamma(u, y) = \min_{\nu, z} \begin{pmatrix} \nu \\ z \end{pmatrix}^T J \begin{pmatrix} \nu \\ z \end{pmatrix}, \tag{2.2.35}$$

where

$$J = \begin{pmatrix} I & 0 \\ 0 & -I \end{pmatrix}, \tag{2.2.36}$$

and $\{\nu, z\}$ satisfy the constraints (2.2.32a) and (2.2.32b) for given $\{u, y\}$. These constraints can be expressed as

$$G_1 \begin{pmatrix} u \\ y \end{pmatrix} = G_2 \begin{pmatrix} \nu, \\ z \end{pmatrix} \tag{2.2.37}$$

where

$$G_1 = \begin{pmatrix} -G_{11} & I \\ -G_{21} & 0 \end{pmatrix}, G_2 = \begin{pmatrix} G_{12} & 0 \\ G_{22} & -I \end{pmatrix}. \tag{2.2.38}$$

Given a matrix A, let $A_\perp$ denote a matrix of maximal full column rank such that $AA_\perp = 0$. That is, $A_\perp$ is a maximal rank full column rank right annihilator. If A has full column rank, then $A_\perp$ does not exist. (In some notation one could then say that $A_\perp$ is the empty matrix.) Note that, in general, $(A_\perp)^T \neq (A^T)_\perp$. We define $A_\perp^T$ to be $(A_\perp)^T$. Using the $\perp$ notation, the solution of the above optimization problem (2.2.35), (2.2.37) can be expressed as follows.

Lemma 2.2.3 *Suppose either that G_2 has full column rank or that*

$$G_{2\perp}^T J G_{2\perp} > 0. \tag{2.2.39}$$

Then the solution to the optimization problem (2.2.35), (2.2.37) is

$$\gamma(u, y) = \begin{pmatrix} u \\ y \end{pmatrix}^T G_1^T (G_2 J G_2^T)^{-1} G_1 \begin{pmatrix} u \\ y \end{pmatrix}. \tag{2.2.40}$$

Proof. Lemma 2.2.3 is a special case of the more general theorem 2.6.1 which will be proved later in this chapter.   □

The assumption (2.2.39) is not restrictive. To see this, suppose that G_2 is not of full column rank and let

$$G_{2\perp} = \begin{pmatrix} \Phi_1 \\ \Phi_2 \end{pmatrix}. \tag{2.2.41}$$

From the definition of G_2 we have that $G_{12}\Phi_1 = 0$ and $G_{22}\Phi_1 - \Phi_2 = 0$. Thus $\Phi_2 = G_{22}\Phi_1$, which implies that

$$G_{2\perp} = \begin{pmatrix} I \\ G_{22} \end{pmatrix} \Phi_1. \tag{2.2.42}$$

But $G_{2\perp}$ has full column rank by construction. But then (2.2.42) implies that Φ_1 has full column rank. It is also straightforward to show that

$$G_{2\perp}^T J G_{2\perp} = \Phi_1^T (I - G_{22}^T G_{22}) \Phi_1, \tag{2.2.43}$$

which means that assumption (2.2.39) of lemma 2.2.3 is satisfied if (2.2.31) holds.

The function γ of (2.2.40) provides the realizability test. The test may be written more compactly. For every symmetric matrix S, there is an invertible matrix K so that $K^T S K$ is a diagonal matrix with diagonal entries of $1, -1,$ or 0. Thus there are $H, \bar{J}$ such that

$$H^T \bar{J} H = G_1^T (G_2 J G_2^T)^{-1} G_1, \tag{2.2.44}$$

where $\bar{J}$ is a signature matrix (diagonal matrix with $+1$ and -1 on the diagonal). If we let

$$\mu = H \begin{pmatrix} u \\ y \end{pmatrix}, \tag{2.2.45}$$

then the test for realizability that we are looking for consists simply in verifying that

$$\mu^T \bar{J} \mu \leq 0. \tag{2.2.46}$$

Example 2.2.2 *Consider the following system:*

$$y = (1 + \Delta) u \tag{2.2.47}$$

with $|\Delta| \leq 1$. *From (2.2.30), it is easy to see that we can take* $G_{22} = 0$ *and* $G_{11} = G_{12} = G_{21} = 1$. *Thus*

$$G_1 = \begin{pmatrix} -1 & 1 \\ -1 & 0 \end{pmatrix}, \quad G_2 = \begin{pmatrix} 1 & 0 \\ 0 & -1 \end{pmatrix}. \tag{2.2.48}$$

It is easy to verify that

$$(G_2 J G_2^T)^{-1} = J \tag{2.2.49}$$

so that we can let $H = G_1$ *in (2.2.44), that is,*

$$\mu = \begin{pmatrix} y - u \\ -u \end{pmatrix}. \tag{2.2.50}$$

The realizability test then becomes

$$\mu^T J \mu = y^2 - 2uy \leq 0. \tag{2.2.51}$$

To see why this inequality is correct, simply note that $y^2 - 2uy = u^2(\Delta^2 - 1)$.
  This inequality defines the set of realizable $\{u, y\}$. *See figure 2.2.4 for an illustration of this set.*

Note that this modeling of uncertainty, as can be seen from figure 2.2.4, for the above example, corresponds to an uncertainty on the slope of the line relating y to u. This is to be compared to figure 2.2.2 where the uncertainty is not in the slope but in the origin. In many cases, neither of these two models is satisfactory. What is needed is an uncertainty model that can capture both of the above uncertainties.

This can be done by relaxing the constraint (2.2.34) to be $|\nu|^2 - |z|^2 < d$ for some positive d. Then the uncertain model is expressed as

$$G_1 \begin{pmatrix} u \\ y \end{pmatrix} = G_2 \begin{pmatrix} \nu \\ z \end{pmatrix} \tag{2.2.52}$$

with the constraint

$$\begin{pmatrix} \nu \\ z \end{pmatrix}^T J \begin{pmatrix} \nu \\ z \end{pmatrix} < d, \tag{2.2.53}$$

where J is defined in (2.2.36). This model captures many natural system uncertainties.

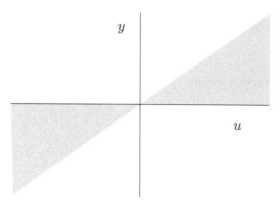

Figure 2.2.4 The realizable set $\{u, y\}$.

Example 2.2.3 *Consider the following system:*

$$\begin{pmatrix} -1 & 1 \\ -1 & 0 \end{pmatrix} \begin{pmatrix} u \\ y \end{pmatrix} = \begin{pmatrix} 1 & 0 \\ 0 & -1 \end{pmatrix} \begin{pmatrix} \nu_1 \\ \nu_2 \end{pmatrix} \qquad (2.2.54)$$

with

$$\begin{pmatrix} \nu_1 \\ \nu_2 \end{pmatrix}^T \begin{pmatrix} 1 & 0 \\ 0 & -1 \end{pmatrix} \begin{pmatrix} \nu_1 \\ \nu_2 \end{pmatrix} < 1. \qquad (2.2.55)$$

This system is closely related to the one studied in example 2.2.2. The realizable set is defined by

$$y^2 - 2uy < 1. \qquad (2.2.56)$$

Figure 2.2.5 illustrates this set of realizable $\{u, y\}$.

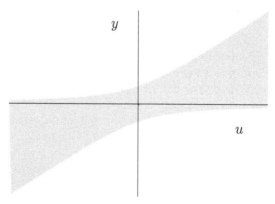

Figure 2.2.5 The realizable set $\{u, y\}$.

Suppose that u is fixed and we are looking at the possible values of y. Looking at figures 2.2.2, 2.2.4, and 2.2.5, we see a major difference between the additive uncertainty and the model uncertainty cases. In figure 2.2.2, as u varies, the set of realizable outputs y stays the same size but is translated. In figures 2.2.4 and 2.2.5, we see that not only may the set of outputs be translated but it can change in size as u varies. This difference will become important later in this book when we study multimodel identification.

2.3 CONTINUOUS-TIME SYSTEMS

In this section we consider again the different types of perturbations we studied in the previous section, but in a more general setting. We are interested in each case in finite-dimensional linear systems of differential equations over the continuous time interval $[0, T]$. We let t be the independent variable but will usually not include it in formulas unless it is necssary to avoid confusion in using the formulas. Some approaches, such as those that involve monitoring an ongoing process, are formulated on an infinite interval $[0, \infty)$. In order to help put our approach in perspective, we shall also make some comments relating our finite-interval approach to the infinite-interval case.

2.3.1 No Uncertainty

If there is no uncertainty, we have the process P completely characterized in terms of the linear system G which is given by

$$\dot{x} = Ax + Bu, \tag{2.3.1a}$$
$$y = Cx + Du. \tag{2.3.1b}$$

Here $\dot{x} = dx/dt$ and x, y, u, A, B, C, D are continuous functions of t. Note that this dependence on t in A, B, C, D can, in fact, be through u. For example, we could have $A(t, u)$, because u is measured on-line so that $u(t)$ is available at time t, and for all practical purposes $A(t, u)$ can be considered just a function of t since the development of the test assumes that y, u are known. This arises frequently in some types of applications. For example, Bu may be in the form of $B(u_1)u_2$, where u_2 is a thrust vector and u_1 are steering angles.

Later, in chapter 3, we will have an additional auxiliary signal v. It can enter (2.3.1a) in the form of $\bar{B}(u, t)v$ where u is known. The techniques of chapter 4 can handle problems of the more general form $B(u, v, t)$.

In theory at least, it is possible to let the residual generator H (which is now also a linear dynamical system) be defined as follows:

$$\dot{\hat{x}} = A\hat{x} + Bu, \tag{2.3.2a}$$
$$\mu = C\hat{x} + Du - y. \tag{2.3.2b}$$

Suppose that $x(0)$ is known. If the initial condition $\hat{x}(0)$ is taken to be equal to $x(0)$, then x is identically equal to $\hat{x}$ and μ is identically zero. Thus the set of $\{u, y\}$ corresponding to normal operation of the system is given by those $\{u, y\}$'s for which μ, given by (2.3.2) with $\hat{x}(0) = x(0)$, is identically zero.

However, in some cases, the slightest discrepancy between $x(0)$ and $\hat{x}(0)$ could result in the divergence of the estimation error

$$\tilde{x} = x - \hat{x} \tag{2.3.3}$$

since it satisfies the differential equation

$$\dot{\tilde{x}} = A\tilde{x}. \tag{2.3.4}$$

This could result in the divergence of the residual μ. In the time-invariant case, this happens when A is not Hurwitz (has eigenvalues with a positive real part) and the time interval is long or infinite.

It is for this reason that on long intervals (2.3.2a) is usually replaced with an *observer* equation:

$$\dot{\hat{x}} = A\hat{x} + Bu + L(C\hat{x} + Du - y). \tag{2.3.5}$$

It is straightforward to verify that now

$$\dot{\tilde{x}} = (A + LC)\tilde{x}, \tag{2.3.6}$$

so that the matrix L must be chosen in such a way that the system (2.3.6) is asymptotically stable.

With an observer, even if the initial condition $\hat{x}$ does not correspond to the actual initial state, the residual μ converges to zero. Of course, if no a priori bound is considered on $x(0)$, the convergence can take an arbitrarily long time to bring μ to a reasonably small value.

Stationary Case

For practical purposes, in the context of failure detection, one always wants to do failure detection and model identification over finite time intervals. However, it is known that a number of time-varying control problems admit simpler algebraic solutions when time-invariant systems $(A, B, C, D$ constant) are considered over an infinite horizon, that is, $[0, \infty)$. The infinite-horizon case is not only of theoretical interest but can often be used to get solutions which are useful over moderate time intervals.

In the time-invariant case, asymptotic stability of system (2.3.5) implies in particular that L must be chosen such that $A + LC$ be Hurwitz. Such an L exists if the pair (C, A) is detectable, that is, all of its unstable modes are observable, or equivalently

$$\begin{pmatrix} sI - A \\ C \end{pmatrix}$$

has full column rank over the complex numbers for all s in the closed right hand plane of $\mathbb{C}$.

The stability property in this case implies that, even if the initial condition $\hat{x}(0)$ is not equal to $x(0)$, then if $\{u, y\}$ correspond to normal operation of the system, μ converges exponentially to zero. Thus, even though here μ does not allow us to completely specify the set of realizable $\{u, y\}$'s without further characterization of the set of possible $x(0)$, it does provide a test. Failure has occurred if μ does not converge to zero exponentially.

Since the choice of $\hat{x}(0)$ is not important when working on an infinite interval, we can set it to 0 and express H in terms of its transfer function. This is done by taking the Laplace transform of equations (2.3.5) and (2.3.2b) and then solving for $\mu(s) = H(s) \begin{pmatrix} u(s) \\ y(s) \end{pmatrix}$ to get

$$H(s) = C(sI - A - LC)^{-1} \left(B + LD \quad -L \right) + \left(D \quad -I \right). \qquad (2.3.7)$$

Once Laplace transforms are taken the system is an algebraic one and appears formally like those studied earlier in the static case. The problem of construction of $H(s)$ can be stated as in (2.2.3) if we replace G by $G(s)$ and T by $T(s)$. Then the $H(s)$ obtained in (2.3.7) is obtained by a particular choice of $T(s)$.

There has been a considerable amount of work using the transfer function to examine failure detection; see, for example, [30]. However, our emphasis in this book is more on detection on short time intervals, so we will not discuss this approach in detail, although we will mention some connections to it as we proceed.

2.3.2 Additive Uncertainty

Suppose now that we have some additive uncertainty in both the dynamics (2.3.1a) and the output equation (2.3.1b). In this case we have

$$\dot{x} = Ax + Bu + M\nu, \qquad (2.3.8a)$$

$$y = Cx + Du + N\nu \qquad (2.3.8b)$$

over the interval $[0, T]$. The uncertainty here comes not only from ν, but also from the initial condition $x(0)$. Having ν in both (2.3.8a) and (2.3.8b) is not restrictive. It is actually more general than the usual case when two different noises are given since the two-noise case is included in (2.3.8) by letting ν contain both noises and using M, N with appropriate block structure. Some information is needed on the type of allowable uncertainty. We consider the following bound on the uncertainty:

$$(x(0) - x_0)^T P_0^{-1}(x(0) - x_0) + \int_0^T |\nu|^2 dt < d, \qquad (2.3.9)$$

where P_0 is a symmetric positive definite matrix. Uncertainty in $x(0)$ is measured with respect to a known x_0. Note that no weighting matrix is specified for ν because, without loss of generality, weighting on ν can be included in the M and N matrices. We do assume, however, that N has full row rank for technical reasons. Note that, rather than restricting the uncertainty, this assumption on N says that all output channels have some noise.

In the estimation context, this type of uncertainty modeling has been studied by Bertsekas and Rhodes [8], who showed that the state of system x based on past observations is an ellipsoid with parameters that are obtained by solving a "Kalman filter" set of equations. The Kalman filter would be the filter obtained for this system if we assumed that ν were a unit-variance zero-mean white-noise process independent of $x(0)$, which is supposed to be a random vector with mean x_0 and covariance P_0.

A priori, the deterministic estimation problem and the Kalman filtering problem are very different in nature. The resemblance in their solutions is due to the fact that in the stochastic setting the left hand side of the inequality (2.3.9) is the log-likelihood of the noise process. In fact, this interpretation of the inequality (2.3.9) is important in the context of failure detection. It states that we consider unlikely noise trajectories, namely, those below the level determined by d, to be outside the set of normal behaviors. Then, if it is unlikely that an observation comes from the model, a failure has occurred. So we can use statistical reasoning to construct the set of normal behaviors used to formulate a simple deterministic problem for solving the failure detection problem.

Statistical reasoning also shows that the inequality (2.3.9) is a reasonable way of bounding the noise process. In some cases it is even superior to instantaneous bounding such as $|\nu(t)| < d$, which may seem appealing in physical systems, where the bounds correspond to physical bounds, but can lead to very conservative formulations.

Noise bounds of the form (2.3.9) will play a fundamental role in several of the later chapters. P_0^{-1} is a positive definite matrix. The inverse appears in order to simplify some of the later formulas. Note that the larger is P_0^{-1}, the more certain we are about $x(0)$ being close to x_0. In some problems, one has no information about x_0, in which case, with a slight abuse of notation, we can take $P_0^{-1} = 0$. This case will be discussed in a later chapter. However, this case requires a more technical discussion, so we do not discuss it further here.

The detection problem is to determine the realizability of $\{u, y\}$, that is, to test if, given the u and y vector functions over the interval $[0, T]$, there exists x such that (2.3.8) and (2.3.9) are satisfied. To solve this problem, we generalize the procedure in the static case and formulate an optimization problem which consists of minimizing the right hand side of (2.3.9),

$$\gamma(u, y) = \min \, (x(0) - x_0)^T P_0^{-1}(x(0) - x_0) + \int_0^T |\nu|^2 dt, \qquad (2.3.10)$$

subject to (2.3.8).

Failure will have occurred if $\gamma(u, y) \geq d$ since that means that too much noise is required to make u, y consistent with the model (2.3.8), or from the stochastic point of view, the most likely consistent noise trajectory is not likely enough.

Theorem 2.3.1 *The solution to the optimization problem (2.3.10), (2.3.8) is given by*

$$\gamma(u, y) = \int_0^T (y - C\hat{x})^T R^{-1}(y - C\hat{x}) \, dt, \qquad (2.3.11)$$

where $\hat{x}$ satisfies the "Kalman filter" equation

$$\dot{\hat{x}} = (A - SR^{-1}C - PC^T R^{-1}C)\hat{x} \\ + (SR^{-1} + PC^T R^{-1})y + Bu, \quad \hat{x}(0) = x_0, \qquad (2.3.12)$$

with P given by the Riccati equation

$$\dot{P} = (A - SR^{-1}C)P + P(A - SR^{-1}C)^T - PC^T R^{-1}CP \\ + Q - SR^{-1}S^T, \quad P(0) = P_0, \qquad (2.3.13)$$

and

$$\begin{pmatrix} Q & S \\ S^T & R \end{pmatrix} = \begin{pmatrix} M \\ N \end{pmatrix} \begin{pmatrix} M \\ N \end{pmatrix}^T. \tag{2.3.14}$$

Proof. Letting $\overline{\Gamma} = I, \overline{a} = Bu, \overline{b} = y - Du, \overline{D} = N$, and $\overline{B} = M$, we see that this problem is a special case of theorem 2.6.2 which is proved later in this chapter. Since $\overline{\Gamma} = I$, the necessary condition (2.6.18) holds. Theorem 2.3.1 then follows immediately from theorem 2.6.2 and corollary 2.6.2. $\square$

If we let H be the following system:

$$\dot{\hat{x}} = (A - SR^{-1}C - PC^T R^{-1}C)\hat{x}$$
$$+(SR^{-1} + PC^T R^{-1})y + Bu, \tag{2.3.15a}$$
$$\mu = R^{-1/2}(y - C\hat{x}) \tag{2.3.15b}$$

with $\hat{x}(0) = x_0$, then the realizability test is reduced to

$$\int_0^T |\mu|^2 \, dt < d. \tag{2.3.16}$$

Note that in general the solution of the Riccati equation P can be computed independently of the measured input and output $\{u, y\}$ unless the system matrices depend on u. In that case, the Riccati equation must be solved on-line along with the filter equation (2.3.15a).

Stochastic Interpretation

The system (2.3.15) corresponds exactly to the Kalman filter for the system (2.3.8) if we interpret ν as a zero-mean, unit-variance white process independent of $x(0)$, a random vector with mean x_0 and covariance P_0. Then μ is the innovation process which is also of zero-mean, unit-variance and white.

These properties of the innovation process can be used to construct statistical failure detection tests. This has been used in particular by Mehra and Peschon [59] for developing a powerful failure detection methodology.

Early Detection

In some cases, it may not be necessary to wait until the end of the test period to decide if a failure has occurred. Noting that the quantity integrated in (2.3.11) is positive, it is easy to see that, if at any time $s \leq T$,

$$\gamma_s(u, y) = \int_0^s (y - C\hat{x})^T R^{-1}(y - C\hat{x}) \, dt \tag{2.3.17}$$

goes above d, a failure can be declared. Note, however, that we cannot rule out failures until $s = T$. Being able to sometimes make a decision before the end of the test period is highly desirable and is a problem that we will address with several of the tests in later chapters.

Stationary Case

In the infinite-horizon, time-invariant case, it is well known that under certain conditions the Riccati equation (2.3.15a) converges and the resulting Kalman filter is asymptotically stable.

Theorem 2.3.2 *Suppose that for all s in the closed right hand plane of $\mathbb{C}$ the matrix*

$$\begin{pmatrix} -sI + A & M \\ C & N \end{pmatrix}$$

has full row rank over the complex numbers and that (C, A) is detectable. Then the Riccati equation (2.3.13) converges exponentially to $P = \bar{P}$, which is the unique positive definite solution of the algebraic Riccati equation

$$0 = (A - SR^{-1}C)\bar{P} + \bar{P}(A - SR^{-1}C)^T - \bar{P}C^T R^{-1}C\bar{P}$$
$$+ Q - SR^{-1}S^T. \qquad (2.3.18)$$

In addition, the system H converges to the asymptotically stable stationary linear system

$$\dot{\hat{x}} = (A - SR^{-1}C - \bar{P}C^T R^{-1}C)\hat{x} + (SR^{-1} + \bar{P}C^T R^{-1})y$$
$$+ Bu, \qquad (2.3.19a)$$
$$\mu = R^{-1/2}(y - C\hat{x}). \qquad (2.3.19b)$$

The filter (2.3.19) can be used instead of (2.3.15) as a suboptimal solution. This is often done when using the Kalman filter. The advantages are considerable in terms of on-line implementation complexity. In particular, there is no Riccati equation to solve on-line, and the filter is time invariant. The loss of optimality can be neglected if T is large compared to the transient response time of the system.

2.3.3 Model Uncertainty

In the static problem when model uncertainty was allowed, the problem became more complex in part because the unconstrained noise bounds were no longer positive semidefinite and one had to carefully consider how the noise bound interacted with the constraints. This occurs again with continuous-time systems when we consider model uncertainty. Suppose that we have a system with uncertain coefficients and uncertain initial conditions given by

$$\dot{x} = (A + M\Delta(I - J\Delta)^{-1}G)x + (B + M\Delta(I - J\Delta)^{-1}H)u, \quad (2.3.20a)$$
$$y = (C + N\Delta(I - J\Delta)^{-1}G)x + (D + N\Delta(I - J\Delta)^{-1}H)u, \quad (2.3.20b)$$
$$\bar{\sigma}(\Delta(t)) \le d_1, \qquad (2.3.20c)$$
$$(x(0) - x_0)^T P_0^{-1}(x(0) - x_0) < d. \qquad (2.3.20d)$$

It is assumed throught this book that Δ is a measurable function. By adjusting the other terms such as H, G, J we may assume that $d_1 = 1$.

Following the static example we suppose that system P of figure 2.1.1 is modeled over the period $[0, T]$ as follows:

$$\dot{x} = Ax + Bu + M\nu, \qquad (2.3.21a)$$
$$z = Gx + Hu + J\nu, \qquad (2.3.21b)$$
$$y = Cx + Du + N\nu, \qquad (2.3.21c)$$

where ν and z are, respectively, the noise input and noise output representing model uncertainty. We assume that N has full row rank. Letting

$$\nu = \Delta z \qquad (2.3.22)$$

we see that (2.3.21) becomes (2.3.20a), (2.3.20b).

Condition (2.3.22) implies that $|\nu(t)| \le |z(t)|$ for all $0 \le t \le T$. Thus we have that

$$\int_0^s (|\nu|^2 - |z|^2) dt < 0, \quad \forall\, s \in [0, T]. \qquad (2.3.23)$$

Combining (2.3.20d) and (2.3.23), we get that the uncertainty on the initial conditions ν and z satisfies

$$(x(0) - x_0)^T P_0^{-1} (x(0) - x_0) + \int_0^s (|\nu|^2 - |z|^2) dt < d, \ \forall\, s \in [0, T]. \qquad (2.3.24)$$

It is important to note that both going to (2.3.23) and then going to (2.3.24) have increased the amount of uncertainty so the answer we shall get is conservative. We discuss this point more carefully later.

The system (2.3.21) along with the noise bound (2.3.24) can also be used to model, with some conservatism, systems where both additive noise and model uncertainties are present. Consider

$$\dot{x} = (A + M\Delta G)x + (B + M\Delta H)u + \bar{B}w, \qquad (2.3.25a)$$

$$y = (C + N\Delta G)x + (D + N\Delta H)u + \bar{D}w, \qquad (2.3.25b)$$

$$\bar{\sigma}(\Delta(t)) \le 1, \qquad (2.3.25c)$$

$$(x(0) - x_0)^T P_0^{-1} (x(0) - x_0) + \int_0^T |w|^2\, dt < d. \qquad (2.3.25d)$$

In this case, we can use system (2.3.21) as follows:

$$\dot{x} = Ax + Bu + \begin{pmatrix} M & \bar{B} \end{pmatrix} \begin{pmatrix} \nu_1 \\ \nu_2 \end{pmatrix}, \qquad (2.3.26a)$$

$$z = Gx + Hu, \qquad (2.3.26b)$$

$$y = Cx + Du + \begin{pmatrix} N & \bar{D} \end{pmatrix} \begin{pmatrix} \nu_1 \\ \nu_2 \end{pmatrix}, \qquad (2.3.26c)$$

with

$$\nu_1 = \Delta z, \quad \nu_2 = w. \qquad (2.3.27)$$

But then we have $|\nu_1(t)| \le |z(t)|$ for all $0 \le t \le T$. Thus we have that

$$\int_0^s (|\nu_1|^2 - |z|^2) < 0, \quad \forall\, s \in [0, T], \qquad (2.3.28)$$

yielding

$$(x(0) - x_0)^T P_0^{-1} (x(0) - x_0) + \int_0^s (|\nu_1|^2 + |\nu_2|^2 - |z|^2) dt < d, \ \forall s \in [0, T] \qquad (2.3.29)$$

which can be rewritten as

$$(x(0) - x_0)^T P_0^{-1}(x(0) - x_0) + \int_0^s \left(\left| \begin{pmatrix} \nu_1 \\ \nu_2 \end{pmatrix} \right|^2 - |z|^2 \right) dt < d, \ \forall \, s \in [0, T],$$

(2.3.30)

which is exactly in the form (2.3.24).

The basic problem that we consider then is system (2.3.21) along with the noise bound (2.3.24). Note that the noise bound (2.3.24) is not equivalent to (2.3.20d) and (2.3.23). Letting $s = 0$ we see that (2.3.24) implies (2.3.20d). However, (2.3.24) does not also imply (2.3.23). This means that any test derived for the problem we consider here may be conservative for the problem (2.3.20a)–(2.3.20d).

Note also that the condition (2.3.24) is stated for all $s \in [0, T]$, while the analogous condition (2.3.9) for additive noises was not. The reason for this difference is that if condition (2.3.9) holds for T it holds for all $s \in [0, T]$, while that is not the case for (2.3.24).

The problem formulation (2.3.21), (2.3.24) is a generalization of the static system (2.2.52), (2.2.53). Here, in addition, the value of d has a nice interpretation in terms of the uncertainty in the initial state.

In the static case of model uncertainty, the realizability test was given in terms of the solution of an optimization problem and, in the nontrivial case, required a matrix product to be nonnegative. The same thing happens here but the solution is somewhat more complex.

Theorem 2.3.3 *For the system (2.3.21) with noise bound (2.3.24) suppose the following:*

1. *Either N is invertible or we have*

$$N_\perp^T (I - J^T J) N_\perp > 0, \quad \forall \, t \in [0, T]$$

(2.3.31)

where $N_\perp$ is a maximal rank full column rank matrix such that $NN_\perp = 0$;

2. *The Riccati equation*

$$\dot{P} = (A - SR^{-1}\bar{C})P + P(A - SR^{-1}\bar{C})^T - P\bar{C}^T R^{-1}\bar{C}P$$
$$+ Q - SR^{-1}S^T, \quad P(0) = P_0,$$

(2.3.32)

where

$$Q = MM^T, \qquad\qquad S = \begin{pmatrix} MN^T & MJ^T \end{pmatrix},$$
$$R = \begin{pmatrix} NN^T & NJ^T \\ JN^T & -I \end{pmatrix}, \qquad \bar{C} = \begin{pmatrix} C \\ G \end{pmatrix},$$

has a solution on $[0, T]$.

Then a realizability test is

$$\gamma_s(u, y) < d, \quad \text{for all } s \in [0, T],$$

(2.3.33)

where

$$\gamma_s(u, y) = \int_0^s \mu^T R^{-1} \mu \, dt$$

(2.3.34)

and μ is the output of the following system:

$$\dot{\hat{x}} = (A - SR^{-1}\bar{C} - P\bar{C}^T R^{-1}\bar{C})\hat{x}$$

$$+ (S + P\bar{C}^T)R^{-1}\begin{pmatrix} y - Du \\ -Hu \end{pmatrix} + Bu, \qquad (2.3.35a)$$

$$\mu = \bar{C}\hat{x} - \begin{pmatrix} y - Du \\ -Hu \end{pmatrix} \qquad (2.3.35b)$$

with $\hat{x}(0) = x_0$.

This result is a generalization of theorem 4.3.1 in chapter 4 of [72].

Proof. We proceed as we did in the static case by defining the following optimization problem:

$$\gamma_s(u, y) = \min_{x(0), \nu} (x(0) - x_0)^T P_0^{-1}(x(0) - x_0) + \int_0^s \begin{pmatrix} \nu \\ z \end{pmatrix}^T \Gamma \begin{pmatrix} \nu \\ z \end{pmatrix} dt, \quad (2.3.36)$$

where $\Gamma = \begin{pmatrix} I & 0 \\ 0 & -I \end{pmatrix}$. But the solution of problem (2.3.36) is a special case of theorem 2.6.2. □

Note that a failure has occurred if $\gamma_s(u, y) \geq d$ for any value of s. Thus this approach has the desirable attribute of being able to sometimes show a failure before the end of the test period T. This will be referred to as early detection and will be discussed more carefully later.

Also note that in general the solution of the Riccati equation P can be computed independently of the measured input and output (u, y) unless the system matrices depend on u. If they depend on u, then the Riccati equation must be solved on-line along with the filter equation (2.3.35a).

As mentioned earlier, by reformulating the uncertain system (2.3.20) into (2.3.21) and (2.3.24), we introduce some conservatism. Indeed, the set of system trajectories allowed under the latter model can be a lot larger than that of the former.

Part of this conservatism was introduced when we obtained (2.3.23) from (2.3.22). This will always be there since some additional $\Delta(t)$ are now permitted. For example, if $\bar{\sigma}(\Delta(t)) < 1$ for a while, then $\bar{\sigma}(\Delta(t)) > 1$ is possible later. We give a specific illustration in the next example.

Some additional conservatism is also introduced when we derived (2.3.24) by adding the terms (2.3.20d) and (2.3.23). For example, if $(x(0) - x_0)^T P_0^{-1}(x(0) - x_0) = c < d$, then we allow Δ for which

$$\int_0^s (|\nu|^2 - |z|^2) \, dt < d - c, \quad \forall \, s \in [0, T].$$

It is easy to show that (2.3.20d) and (2.3.23) are equivalent to

$$(x(0) - x_0)^T P_0^{-1}(x(0) - x_0) + \alpha \int_0^s (|\nu|^2 - |z|^2) \, dt < d, \quad \forall s \in [0, T], \forall \, \alpha > 0,$$

since we could have added any positive multiple of (2.3.23) to (2.3.20d). It turns out that this additional degree of freedom is still available in the model through the choice of P_0 and d. Note that the set of consistent initial states $x(0)$ defined by (2.3.20d) depends only on the product dP_0, and consequently so does the whole uncertain system (2.3.20). The individual choices of d and P_0, however, affect the set of system trajectories defined by (2.3.21) and (2.3.24).

The effect of P_0 and d on the new problem formulation is illustrated in the following example.

Example 2.3.1 *Consider a simple example of the uncertain system (2.3.20):*

$$\dot{x} = (-1 + \delta)x, \quad x(0)^2 < 1, \quad |\delta(t)| < 1, \tag{2.3.37}$$

where there is no output equation (2.3.20b). Again $\delta(t)$ is assumed to be a measurable function. It is not difficult to verify that for any t we have

$$x(t)^2 < 1, \tag{2.3.38}$$

and the constraint (2.3.38) is tight as a pointwise bound. However, not all such functions can be trajectories. For example, it is easy to see that $0 \geq \dfrac{d}{dt}\left(x^2\right) \geq -2x^2$. Thus, in addition to (2.3.38), we have that the possible x^2 are monotonically nonincreasing and there is a lower bound on how fast x can decrease.

To illustrate the conservative nature of our approach, suppose that we set $\delta(t) = 0$ for $0 \leq t \leq 1$. Then a straightforward calculation shows that (2.3.23) holds for $0 \leq s \leq 1$ and for $s > 1$ it becomes

$$\int_1^s e^{2\int_0^\tau -1+\delta(r)dr}d\tau \leq \int_0^1 e^{-2\tau}d\tau. \tag{2.3.39}$$

Clearly, this allows for $\delta(t) > 1$ for some $t > 1$.

We now turn to illustrating the earlier point about the degree of freedom in P_0, d. Proceeding as proposed in this section, we convert this problem into

$$\dot{x} = -x + \nu, \tag{2.3.40}$$

$$z = x, \tag{2.3.41}$$

with no input u and output y, and $dP_0 = 1$. This gives us the Riccati equation

$$\dot{P} = (P - 1)^2, \quad P(0) = P_0. \tag{2.3.42}$$

In this new formulation, we have the following constraint on $x(t)$:

$$x(t)P(t)^{-1}x(t) < d. \tag{2.3.43}$$

This implies that

$$x(t)^2 < dP(t), \tag{2.3.44}$$

where $P(t)$ satisfies

$$\dot{P} = (P - 1)^2, \quad P(0) = 1/d. \tag{2.3.45}$$

In this case, we can explicitly solve the Riccati equation for P and obtain a closed-from expression:

$$x(t)^2 < d\frac{(d-1)t + 1}{(d-1)t + d}. \tag{2.3.46}$$

Clearly, for $d = 1$, we get exactly condition (2.3.38). We do not have any extra conservatism beyond that given by using (2.3.23) instead of (2.3.22). For $d > 1$, the bound (2.3.46) becomes bigger as t increases. The limit as t goes to infinity gives us a bound of d instead of 1. For the case $d < 1$, the Riccati equation diverges at $t = d/(1 - d)$.

We have seen in the above example that the choice of the pair (d, P_0) is important in obtaining a good approximation when reformulating the uncertain system. Note that if $\overline{\sigma}(J) < 1$, the system (2.3.20a) is a linear time-varying system with bounded coefficients over $[0, T]$, so its solution is always bounded. Thus it would be reasonable, in general, to choose the pair (d, P_0), if possible, in such a way that the Riccati equation associated with the new formulation does not diverge, no matter what the output equation is. It turns out that the worst output equation is the absence of an output equation.

Lemma 2.3.1 *Suppose*

1. $\overline{\sigma}(J) < 1$,

2. The Riccati equation
$$\dot{P} = (A + MJ^TG)P + P(A + MJ^TG)^T + PG^TGP$$
$$+ M(I + J^TJ)M^T, \quad P(0) = P_0, \tag{2.3.47}$$
has a bounded solution on $[0, T]$.

Then the conditions of theorem 2.3.3 are satisfied. In addition, the state trajectories x form a convex, pointwise bounded set of functions.

This lemma is a special case of theorem 2.6.3.

The assumptions of lemma 2.3.1 imply the existence of a unique bounded solution to (2.3.36) subject to system (2.3.21) but without the output y. This information is useful in another way. In the next chapter, which considers more than one model, we will be able to design a test, which we call the hyperplane test, when the output sets $\mathcal{A}(u)$ are bounded and convex. In general, the property of being bounded and convex depends on the choice of the system matrices in the output equations. An important special case is when the output set is bounded and convex for all output functions. This happens when the set of all possible state trajectories is a convex set which is bounded in L^2. From Lemma 2.3.1 we get the following result which we shall use in the next chapter.

Lemma 2.3.2 *Suppose the conditions of lemma 2.3.1 are satisfied. Then the output set $\mathcal{A}(u)$, the set of y consistent with (2.3.21) and (2.3.24), is bounded in L^2 and is convex.*

Note that the conditions of lemma 2.3.1 are sufficient but not necessary for carrying out the procedures of this chapter. We do not need to assume these conditions. The more permissive conditions of theorem 2.3.3 are enough to construct the detection filter. Similarly, in the next chapter, we propose a method to construct auxiliary signals for multimodel identification under weaker conditions than those of lemma 2.3.1. We do need these assumptions when we propose efficient on-line detection techniques that require convexity of the output sets.

2.4 DISCRETE-TIME SYSTEMS

In many physical systems events occur, or measurements are taken, at discrete times. Even many continuous control algorithms when actually implemented become discrete. Again, we are mainly interested in the finite-interval case but will make some comments about the infinite-interval case.

2.4.1 No Uncertainty

If there is no uncertainty we have the process P completely characterized in terms of the linear system G which is given by

$$x_{k+1} = Ax_k + Bu_k, \tag{2.4.1a}$$

$$y_k = Cx_k + Du_k, \tag{2.4.1b}$$

and k is an integer variable. The coefficients A, B, C, D may also depend on k (possibly through u). For readability we suppress this k dependence unless it is essential. In general, given a finite or infinite sequence $\{z_k\}$ we shall let z denote the entire sequence.

If k runs over a finite set $\{0, 1, \ldots, N-1\}$, then in principle one could rewrite (2.4.1) as a static system:

$$\begin{pmatrix} x_1 \\ \vdots \\ x_N \end{pmatrix} = \begin{pmatrix} A & \cdot & 0 \\ \vdots & \ddots & \vdots \\ 0 & \cdot & A \end{pmatrix} \begin{pmatrix} x_0 \\ \vdots \\ x_{N-1} \end{pmatrix} + \begin{pmatrix} B & \cdot & 0 \\ \vdots & \ddots & \vdots \\ 0 & \cdot & B \end{pmatrix} \begin{pmatrix} u_0 \\ \vdots \\ u_{N-1} \end{pmatrix} \tag{2.4.2a}$$

$$\begin{pmatrix} y_0 \\ \vdots \\ y_{N-1} \end{pmatrix} = \begin{pmatrix} C & \cdot & 0 \\ \vdots & \ddots & \vdots \\ 0 & \cdot & C \end{pmatrix} \begin{pmatrix} x_0 \\ \vdots \\ x_{N-1} \end{pmatrix} + \begin{pmatrix} D & \cdot & 0 \\ \vdots & \ddots & \vdots \\ 0 & \cdot & D \end{pmatrix} \begin{pmatrix} u_0 \\ \vdots \\ u_{N-1} \end{pmatrix} \tag{2.4.2b}$$

Static problems were considered in section 2.2. Thus, in one theoretical sense, the results for the static case already cover discrete systems on finite intervals.

However, from a practical point of view, merely looking at (2.4.1) as a larger static system throws away all the structure of these equations. For many applications, particularly those that require computations to be done in real time, or close to real time, it is necessary to exploit the recursive structure of the difference equations. Note that, if we were to use the static approach, we would have to wait until the end of the test period before even starting our computations. Thus, while we can often turn to the static case during a theoretical argument, the results and algorithms have more in common with the continuous-time case.

In (2.4.1), y_k and u_k are measured. If x_0 is also known, then there is no uncertainty. One could apply a general static realizability test such as (2.2.12) where μ is obtained from (2.2.10). However, one can also recursively construct μ as follows:

$$\hat{x}_{k+1} = A\hat{x}_k + Bu_k, \tag{2.4.3}$$

$$\mu_k = -y_k + C\hat{x}_k + Du_k, \tag{2.4.4}$$

or equivalently, if A, B are constant matrices,

$$\mu_k = C\left(A^k x_0 + \sum_{i=0}^{k-1} A^i B u_{k-1-i}\right) + D u_k - y_k. \tag{2.4.5}$$

Then $\{u, y\}$ is realizable if

$$\mu_k = 0, \quad k = 0, \dots, N-1, \text{ or equivalently } \sum_{i=0}^{N-1} \mu_i^2 = 0. \tag{2.4.6}$$

With this approach, there is no reason to wait until the end of a finite test period to start the detection process. We can process u_k and y_k as they become available. Not only does this mean more efficient computations and reduced memory requirements since there is no need to store u_k and y_k, but it also means that failure can be declared as soon as $\mu_k \neq 0$.

Suppose that the interval is either long or infinite. The error equation governing the estimation error

$$\tilde{x}_k = x_k - \hat{x}_k, \tag{2.4.7}$$

which is given by

$$\tilde{x}_{k+1} = A\tilde{x}_k, \tag{2.4.8}$$

may not converge if $\tilde{x}_0 = \hat{x}_0 - x_0$ is not exactly zero and if (2.4.8) is not asymptotically stable. So, assuming we are in the time-invariant case, this approach works as long as $\rho(A) < 1$. That is, the spectral radius, or equivalently the maximum of all absolute values of eigenvalues of A, is less than 1.

It is for this reason that an observer,

$$\hat{x}_{k+1} = A\hat{x}_k + B u_k - L(y_k - C\hat{x}_k - D u_k), \tag{2.4.9}$$

is often used to construct the estimate $\hat{x}_k$. The error equation then becomes

$$\tilde{x}_{k+1} = (A + LC)\tilde{x}_k, \tag{2.4.10}$$

so that in the time-invariant case, if L is chosen such that $\rho(A + LC) < 1$, then $\tilde{x}_k$ and consequently μ_k converge exponentially to zero. This can be done if

$$\begin{pmatrix} zI - A \\ C \end{pmatrix}$$

has full column rank for all z in $\mathbb{C}$ on and outside the unit circle ($|z| > 1$). If this full rank condition holds for all nonzero z, that is, $\{C, A\}$ is observable, then L can be chosen such that $A + LC$ is nilpotent. This implies that $\tilde{x}_k$ and consequently μ_k converge to zero in finite time. Such observers are called deadbeat and can be used to construct parity checks on the measured quantities u_k and y_k over sliding finite-time windows. Parity check methods have been used for failure detection.

With an observer, even if the initial condition $\hat{x}_0$ does not correspond to the actual initial state, the residual μ_k converges exponentially to zero. Of course, if no a priori bound is considered on $x(0)$, even the exponential convergence can take an arbitrarily long time to bring μ to a reasonably small value, unless a deadbeat observer is used.

2.4.2 Additive Uncertainty

The more important case is when there is an uncertainty in both the dynamics and the output equations. If this uncertainty is due to external disturbances, noise, measurement error, etc., but not to errors in the model itself, then additive noise models are often used.

Suppose then that we have some additive uncertainty in both the dynamics (2.4.3) and the output equation (2.4.4). In this case we have

$$x_{k+1} = Ax_k + Bu_k + M\nu_k, \qquad (2.4.11a)$$
$$y_k = Cx_k + Du_k + N\nu_k \qquad (2.4.11b)$$

for $k = 0, \dots, N - 1$. The uncertainty here comes not only from ν, but also in the initial condition x_0. Some information is needed on the type of allowable uncertainty. We consider the following bound on the uncertainty:

$$(x_0 - \bar{x}_0)^T P_0^{-1}(x_0 - \bar{x}_0) + \sum_{i=0}^{N-1} |\nu_i|^2 < d, \qquad (2.4.12)$$

where P_0 is a symmetric positive definite matrix. This is the discrete counterpart of (2.3.10). Note that again no weighting matrix is specified for the ν_k because, without loss of generality, weighting on ν can be included in the M and N matrices. We do assume, however, that N has full row rank for every k for technical reasons. Rather than being restrictive, this assumption says that all output channels have some noise.

The detection problem of determining the realizability of $\{u, y\}$ is approached in the same manner as with the continuous-time problem. We formulate an optimization problem which consists of minimizing

$$\gamma(u, y) = \min \left\{ (x_0 - \bar{x}_0)^T P_0^{-1}(x_0 - \bar{x}_0) + \sum_{i=0}^{N-1} |\nu_i|^2 \right\} \qquad (2.4.13)$$

subject to (2.4.11).

Failure will have occurred if $\gamma(u, y) \geq d$, since that means that too much noise is required to make u, y consistent with the model (2.4.11). The solution of this optimization problem is given by the following theorem.

Theorem 2.4.1 *The solution to the optimization problem (2.4.13) is given by*

$$\gamma(u, y) = \sum_{k=0}^{N-1} \hat{\mu}_k^T \Delta_k^{-1} \hat{\mu}_k, \qquad (2.4.14)$$

where

$$\Delta_k = CP_k C^T + NN^T, \qquad (2.4.15)$$

$$\hat{\mu}_k = C\hat{x}_k - (y_k - Du_k), \qquad (2.4.16)$$

and $\hat{x}$ satisfies the "Kalman filter" equation

$$\hat{x}_{k+1} = A\hat{x}_k + Du_k - (AP_k C^T + MN^T)(CP_k C^T + NN^T)^{-1}\hat{\mu}_k, \quad \hat{x}_0 = \bar{x}_0, \qquad (2.4.17)$$

where P_k is a positive solution on $[1, N-1]$ of the Riccati equation

$$P_{k+1} = (AP_kA^T + M\Gamma^{-1}M^T)$$
$$- (AP_kC^T + MN^T)(CP_kC^T + NN^T)^{-1}(AP_kC^T + MN^T)^T. \quad (2.4.18)$$

Proof. Theorem 2.4.1 is a special case of the more general theorem 2.6.4 which is proved later in this chapter. Theorem 2.4.1 follows from theorem 2.6.4 by noting that $\Gamma = I$ and condition (2.6.18) holds automatically. $\qquad\square$

2.4.3 Model Uncertainty

Model uncertainty can be incorporated into the discrete-time models much as it was incorporated in the last section for continuous-time models. Suppose that we have a system with uncertain coefficients and uncertain initial conditions given by

$$x_{k+1} = (A + M\Delta_k(I - J\Delta_k)^{-1}G)x_k + (B + M\Delta_k(I - J\Delta_k)^{-1}H)u_k, (2.4.19a)$$
$$y_k = (C + N\Delta_k(I - J\Delta_k)^{-1}G)x_k + (D + N\Delta_k(I - J\Delta_k)^{-1}H)u_k, \quad (2.4.19b)$$
$$\bar{\sigma}(\Delta_k) \leq d_1, \quad (2.4.19c)$$
$$(x_0 - \bar{x}_0)^T P_0^{-1}(x_0 - \bar{x}_0) < d. \quad (2.4.19d)$$

By adjusting the other terms such as H, G, J we may assume that $d_1 = 1$.

Following the static example, we suppose that the system we are interested in is modeled over the period $[0, N-1]$ as follows:

$$x_{k+1} = Ax_k + Bu_k + M\nu_k, \quad (2.4.20a)$$
$$z_k = Gx_k + Hu_k + J\nu_k, \quad (2.4.20b)$$
$$y_k = Cx_k + Du_k + N\nu_k, \quad (2.4.20c)$$

where ν and z are, respectively, the noise input and noise output representing model uncertainty. We assume that N has full row rank for every k. Letting

$$\nu_k = \Delta_k z_k, \quad (2.4.21)$$

we see that (2.4.20) becomes (2.4.19a), (2.4.19b).

The condition (2.4.21) implies that $|\nu_k| \leq |z_k|$ for all $0 \leq k \leq N-1$. Thus we have that

$$\sum_{i=0}^{k}(|\nu_i|^2 - |z_i|^2) < 0, \quad \forall\, k \in [0, N-1]. \quad (2.4.22)$$

Combining (2.4.19d) and (2.4.22) we get the uncertainty on the initial state ν_k, and z_k satisfies

$$(x_0 - \bar{x}_0)^T P_0^{-1}(x_0 - \bar{x}_0) + \sum_{i=0}^{k}(|\nu_i|^2 - |z_i|^2) < d, \forall\, k \in [0, N-1]. \quad (2.4.23)$$

The system (2.4.20) along with the noise bound (2.4.23) can also be used to model (with some conservatism) systems where both additive noise and model uncertainties are present. Consider

$$x_{k+1} = (A + M\Delta_k G)x_k + (B + M\Delta_k H)u_k + \bar{B}w_k, \qquad (2.4.24a)$$

$$y_k = (C + N\Delta_k G)x_k + (D + N\Delta_k H)u_k + \bar{D}w_k, \qquad (2.4.24b)$$

$$\bar{\sigma}(\Delta_k) \le 1, \qquad (2.4.24c)$$

$$(x_0 - \bar{x}_0)^T P_0^{-1}(x_0 - \bar{x}_0) + \sum_{i=0}^{N-1} |w_i|^2 < d. \qquad (2.4.24d)$$

In this case, we can use system (2.4.20) as follows:

$$x_{k+1} = Ax_k + Bu_k + \begin{pmatrix} M & \bar{B} \end{pmatrix} \begin{pmatrix} \nu_1 \\ \nu_2 \end{pmatrix}_k, \qquad (2.4.25a)$$

$$z_k = Gx_k + Hu_k, \qquad (2.4.25b)$$

$$y_k = Cx_k + Du_k + \begin{pmatrix} N & \bar{D} \end{pmatrix} \begin{pmatrix} \nu_1 \\ \nu_2 \end{pmatrix}_k, \qquad (2.4.25c)$$

with

$$\nu_{1k} = \Delta_k z_k, \quad w_k = \nu_{2k}. \qquad (2.4.26)$$

But then we have $|\nu_{1k}| \le |z_k|$. Thus we have that

$$\sum_{i=0}^{k} (|\nu_{1i}|^2 - |z_i|^2) < 0, \qquad (2.4.27)$$

yielding

$$(x_0 - \bar{x}_0)^T P_0^{-1}(x_0 - \bar{x}_0) + \sum_{i=0}^{k} \left(\left| \begin{pmatrix} \nu_1 \\ \nu_2 \end{pmatrix}_i \right|^2 - |z_i|^2 \right) < d, \quad \forall k \in [0, N-1], \qquad (2.4.28)$$

which is exactly in the form (2.4.23).

The basic problem that we will consider then is system (2.4.20) along with the noise bound (2.4.23). Note that the noise bound (2.4.23) is not equivalent to (2.4.19d) and (2.4.22). Letting $k = 0$ we see that (2.4.23) implies (2.4.19d). However, (2.4.23) does not also imply (2.4.22). This means that any test derived for the problem we consider here may be conservative for the problem (2.4.19a)–(2.4.19d).

Note also that one condition (2.4.23) is stated for all $k \in [0, N-1]$, while the analogous condition (2.4.12) for additive noises was not. The explanation is that if condition (2.4.12) holds for $N-1$ it holds for all $k \in [0, N-1]$, while that is not the case for (2.4.23).

The problem formulation (2.4.20),(2.4.23) is a counterpart of the continuous-time system (2.3.21), (2.3.24). Here too the value of d has a nice interpretation in terms of the uncertainty in the initial condition. As with the model uncertain static and continuous-time cases, the realizability test is given in terms of the solution of an optimization problem and requires a matrix product to be nonnegative.

The realizability test consists of letting

$$\gamma_k(u, y) = \min \left\{ (x_0 - \bar{x}_0)^T P_0^{-1}(x_0 - \bar{x}_0) + \sum_{i=0}^{k} (|\nu_i|^2 - |z_i|^2) \right\}, \qquad (2.4.29)$$

subject to the constraints (2.4.20). Then failure occurs if $\gamma_k(u,y) \geq d$.

The next result explains how to compute $\gamma(u,y)$ given u,y. If y_k, u_k are being computed on-line, then the recursive nature of the formulas allows one to compute γ as the data are received.

Theorem 2.4.2 *The optimization problem (2.4.29) with constraints (2.4.20) has a unique bounded solution (x, ν, z) for all $k \in [0, N-1]$ if and only if*

1. *Either* $\begin{pmatrix} A & \tilde{M} \\ \tilde{C} & \tilde{N} \end{pmatrix}$ *is invertible or*

$$\begin{pmatrix} A & \tilde{M} \\ \tilde{C} & \tilde{N} \end{pmatrix}_{\perp}^{T} \begin{pmatrix} P_k^{-1} & 0 \\ 0 & \Gamma \end{pmatrix} \begin{pmatrix} A & \tilde{M} \\ \tilde{C} & \tilde{N} \end{pmatrix}_{\perp} > 0, \quad \forall k \in [0, N-1]; \quad (2.4.30)$$

2. *The Riccati equation*

$$P_{k+1} = (AP_k A^T + MM^T) - \\ (AP_k \tilde{C}^T + \tilde{M}\tilde{N}^T)(\tilde{C}P_k \tilde{C}^T + \tilde{N}\Gamma^{-1}\tilde{N}^T)^{-1}(AP_k \tilde{C}^T + \tilde{M}\tilde{N}^T)^T \quad (2.4.31)$$

has a positive solution P_k on $[1, N]$,

where

$$\Gamma = \begin{pmatrix} I & 0 \\ 0 & -I \end{pmatrix}, \ \tilde{M} = (M \quad 0), \ \tilde{N} = \begin{pmatrix} J & -I \\ N & 0 \end{pmatrix}, \ \tilde{C} = \begin{pmatrix} G \\ C \end{pmatrix}. \quad (2.4.32)$$

Corollary 2.4.1 *The solution to the optimization problem (2.6.15) is given by*

$$\gamma(u,y) = \sum_{k=0}^{j} \hat{\mu}_k^T \Delta_k^{-1} \hat{\mu}_k, \quad (2.4.33)$$

where

$$\Delta_k = \tilde{C}P_k\tilde{C}^T + \tilde{N}\Gamma^{-1}\tilde{N}^T \quad (2.4.34)$$

and

$$\hat{\mu}_k = \tilde{C}\hat{x}_k - \begin{pmatrix} -Hu_k \\ y_k - Du_k \end{pmatrix}, \quad (2.4.35)$$

and where $\hat{x}$ satisfies the "Kalman filter" equation

$$\hat{x}_{k+1} = A\hat{x}_k + Bu_k - (AP_k\tilde{C}^T + \tilde{M}\Gamma^{-1}\tilde{N}^T)(\tilde{C}P_k\tilde{C}^T + \tilde{N}\Gamma^{-1}\tilde{N}^T)^{-1}\hat{\mu}_k, \\ \hat{x}_0 = \bar{x}_0 \quad (2.4.36)$$

with P_k the solution of the Riccati equation in theorem 2.4.2.

Proof. We rewrite (2.4.20) as

$$x_{k+1} = Ax_k + (M \quad 0)\begin{pmatrix} \nu_k \\ z_k \end{pmatrix} + Bu_k, \quad (2.4.37)$$

$$\begin{pmatrix} -Hu_k \\ y_k - Du_k \end{pmatrix} = \begin{pmatrix} G \\ C \end{pmatrix}x_k + \begin{pmatrix} J & -I \\ N & 0 \end{pmatrix}\begin{pmatrix} \nu_k \\ z_k \end{pmatrix}. \quad (2.4.38)$$

This is in the form to apply theorem 2.6.4 if we take

$$w_k = \begin{pmatrix} \nu_k \\ z_k \end{pmatrix}, \ a_k = Bu_k, \ b_k = \begin{pmatrix} -Hu_k \\ y_k - Du_k \end{pmatrix}. \quad (2.4.39)$$

Note that $\Gamma^{-1} = \Gamma$ and $\tilde{M}\Gamma = \tilde{M}$. Making the appropriate substitutions into theorem 2.6.4 now gives theorem 2.4.2. □

2.5 REAL-TIME IMPLEMENTATION ISSUES

There are a number of issues that must be considered when implementing the realizability test developed in this chapter for failure detection in dynamical systems. The most important issue is the conservatism introduced when, for example, in the continuous-time case, condition (2.3.24) is obtained from (2.3.20d) and the fact that $|\nu(t)| \le |z(t)|$, which is implied by (2.3.22) and (2.3.20c). Clearly, using the same arguments, instead of (2.3.24), we could have used a more general expression:

$$(x(0) - x_0)^T P_0^{-1} (x(0) - x_0) + \int_0^s \phi(t)(|\nu|^2 - |z|^2) dt < d, \ \forall \, s \in [0, T],$$

(2.5.1)

where $\phi(t)$ is any function satisfying

$$\phi(t) \ge 0, \ \forall \, t \in [0, s].$$

(2.5.2)

The choice made in this chapter was $\phi(t) = 1$, which resulted in the condition (2.3.24). This is a reasonable choice in the situation where we assume that no failure can occur during the test period. Either the system has failed already, or if not, it remains unfailed until the end. This is the situation that we will consider in the next chapter, where our objective is to design failure detection tests implemented over short time periods. Such tests would then be used either on a regular basis or at critical times. Since each test period is short, it would be reasonable to assume that the failure does not occur during the test. And even if it does, disturbing the test result, the failure would be detected by the next test. That is why we consider only the case $\phi(t) = 1$ for the auxiliary signal design problem which is the main contribution of this book, and is the subject of the following chapters.

The situation would be different if we were to use this test over a long period of time for detecting failures that could come up at any time. That is, the test is being used as part of a continuous monitoring of the system. The averaging effect introduced by the integration over the interval $[0, s]$ in (2.5.1), where s represents the current time, reduces the efficiency of the test. In particular, the longer the test has been running, the worse is the detection. Clearly, this shows serious limitations in the use of this detection test in this situation.

One way to get around this problem, besides frequent restarts, is to use a different $\phi(t)$, and in particular one that weighs recent observations more heavily while forgetting progressively older observations. Different weightings can be used. To illustrate the effect of weightings we will consider one commonly used weight, which is

$$\phi(t) = \exp(\alpha t), \ \alpha > 0,$$

(2.5.3)

which yields the following constraint:

$$(x(0) - x_0)^T P_0^{-1} (x(0) - x_0) + \int_0^s \exp(\alpha t)(|\nu|^2 - |z|^2) \, dt < d, \ \forall \, s \in [0, T].$$

(2.5.4)

The realizability test that is obtained from minimizing the left hand side of (2.5.4) subject to system constraints can still be obtained directly from the application of theorem 2.6.2 and its corollary because in these results all the matrices can be time dependent. The result after some straightforward algebra is the following

Theorem 2.5.1 *For the system (2.3.21) with noise bound (2.5.4) and $\alpha > 0$, suppose that*

1. *Either N is invertible or*
$$N_\perp^T(I - J^T J)N_\perp > 0, \quad \forall t \in [0, T]; \tag{2.5.5}$$

2. *The Riccati equation*
$$\dot{P} = \alpha P + (A - SR^{-1}\bar{C})P + P(A - SR^{-1}\bar{C})^T$$
$$-P\bar{C}^T R^{-1}\bar{C}P + Q - SR^{-1}S^T, \quad P(0) = P_0, \tag{2.5.6}$$

where
$$Q = MM^T, \qquad\qquad S = \left(MN^T \quad MJ^T\right),$$
$$R = \begin{pmatrix} NN^T & NJ^T \\ JN^T & -I \end{pmatrix}, \qquad \bar{C} = \begin{pmatrix} C \\ G \end{pmatrix}$$

has a solution on $[0, T]$.

Then a realizability test is
$$\gamma(s) < d\exp(-\alpha s), \quad \text{for all } s \in [0, T], \tag{2.5.7}$$

where γ satisfies
$$\dot{\gamma} = -\alpha\gamma + \mu^T R^{-1}\mu, \quad \gamma(0) = 0, \tag{2.5.8}$$

and μ is the output of the following system:
$$\dot{\hat{x}} = (A - SR^{-1}\bar{C} - P\bar{C}^T R^{-1}\bar{C})\hat{x} + (S + P\bar{C}^T)R^{-1}\begin{pmatrix} y - Du \\ -Hu \end{pmatrix} + Bu, \tag{2.5.9}$$

$$\mu = \bar{C}\hat{x} - \begin{pmatrix} y - Du \\ -Hu \end{pmatrix} \tag{2.5.10}$$

with $\hat{x}(0) = x_0$.

The result for a typical situation is illustrated in figure 2.5.1. Here we have a randomly generated stable linear system of the form (2.3.20a)–(2.3.20d) with $d_1 = d = 1$ and $J = 0$. The system has two inputs, one constant and the other sinusoidal. For the simulation Δ is set to zero on $[0, 5]$ and then to a constant matrix with norm larger than 1, emulating a failure.

The realizability test consists of comparing $\gamma(s) - d\exp(-\alpha s)$ with zero. Failure is declared as soon as this value becomes positive. Figure 2.5.1 shows this value for different values of α. The $\alpha = 1$ line is dotted to distinguish it from the others. The failure time is illustrated with an upward arrow whereas the detection times for various α's are illustrated by downward arrows.

Note that for large α ($\alpha = 1$) the detection is almost immediate but the test is noise sensitive because $\gamma(s) - d\exp(-\alpha s)$ remains close to zero. This means that the threshold must be chosen in a narrow band. On the other hand, for $\alpha = 0$, which is the standard test, there is a large detection delay, but the test is not noise sensitive. Finally, $\alpha = 0.1$ shows a reasonable compromise. There is a reasonable delay and also reasonable noise sensitivity.

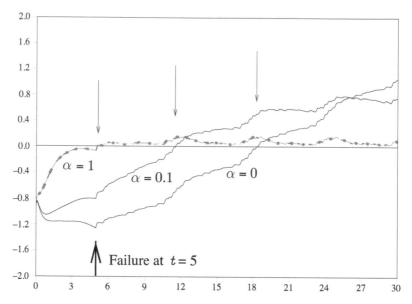

Figure 2.5.1 Plot of $\gamma(s) - d\exp(-\alpha s)$ against time for various values of α.

2.6 USEFUL RESULTS

The previous discussion has shown that for a variety of detection problems the design of a residual, or realizability test, can be reduced to finding the solution of an optimization problem. This pattern will persist in later chapters. In this section we will present several useful results. Some of these have been referred to earlier. All of them will play a role in the developments that follow. Results that are related to a particular previous section are grouped together.

2.6.1 Static Optimization Problem

We have seen that the optimization problem

$$J(a, c) = \min_{\nu}(\nu - c)^T \Gamma(\nu - c) \tag{2.6.1}$$

subject to

$$a = A\nu, \tag{2.6.2}$$

where Γ is an invertible symmetric (but not necessarily positive) matrix, A is a full row rank matrix, and a and c are vectors with appropriate sizes, played an important role in section 2.2. It will be used again later in this book. The solution is given by the following theorem.

Theorem 2.6.1 *For the optimization problem (2.6.1), (2.6.2), if A is not full column rank, let $A_\perp$ be a full column rank matrix of maximum rank such that $AA_\perp = 0$. Let*

$A_\perp^T = (A_\perp)^T$ *and define*

$$\Delta = A_\perp^T \Gamma A_\perp. \tag{2.6.3}$$

Suppose that a is in the range of A so that (2.6.2) is consistent. Then there are three possibilities for $J(a, c)$.

    1. If A has full column rank or $\Delta > 0$, then

$$J(a, c) = (a - Ac)^T (A\Gamma^{-1}A^T)^{-1}(a - Ac); \tag{2.6.4}$$

    2. If $\Delta \geq 0$ but not > 0, A^r is a right inverse of A, and

$$\ker(\Delta) \subset \ker((a - Ac)^T (A^r)^T \Gamma A_\perp), \tag{2.6.5}$$

    then

$$J(a, c) = (a - Ac)^T (A\Gamma^{-1}A^T)^\dagger (a - Ac); \tag{2.6.6}$$

    3. Otherwise $J(a, c) = -\infty$.

Proof. If A has full column rank, then ν is uniquely determined and the result is immediate. Suppose that A is not of full column rank. Let $w = \nu - c$ and $\hat{a} = a - Ac$. Then (2.6.1), (2.6.2) is reduced to the case when $c = 0$. Thus we may assume that $c = 0$. Let ν_0 satisfy

$$\hat{a} = A\nu_0. \tag{2.6.7}$$

Then for any other ν satisfying (2.6.2) there exists a vector ξ such that

$$\nu = \nu_0 + A_\perp \xi. \tag{2.6.8}$$

Let $Q(\xi) = \nu^T \Gamma \nu$ so that

$$Q(\xi) = \nu_0^T \Gamma \nu_0 + 2\nu_0^T \Gamma A_\perp \xi + \xi^T A_\perp^T \Gamma A_\perp \xi. \tag{2.6.9}$$

If it is not true that $\Delta \geq 0$, then there is a vector ϕ so that $\phi^T \Delta \phi < 0$. Letting $\xi = c\phi$ and taking c large, we see that $Q(\xi)$ can be made negative and the minimum value is $-\infty$. Thus the third condition of theorem 2.6.1 holds.

We may suppose that $\Delta \geq 0$. Suppose that $\Delta > 0$. Then $Q(\xi)$ is a quadratic function with a positive definite quadratic term and the minimum exists. To find the minimum, we form the Hamiltonian

$$H = \nu^T \Gamma \nu + \lambda^T (\hat{a} - A\nu).$$

Differentiating with respect to ν, λ we get

$$\hat{a} - A\nu = 0, \tag{2.6.10}$$
$$2\Gamma\nu - A^T \lambda = 0. \tag{2.6.11}$$

The second equation gives $\nu = \dfrac{1}{2}\Gamma^{-1}A^T\lambda$. Multiplying by A and using (2.6.10) gives

$$\hat{a} = \frac{1}{2}(A\Gamma^{-1}A^T)\lambda. \tag{2.6.12}$$

But we know that a minimum will exist for any $\hat{a}$ so that $(A\Gamma^{-1}A^T)$ is invertible. Solving for λ, substituting into (2.6.11), and then solving for ν gives finally that the minimum occurs at

$$\nu = \Gamma^{-1}A^T(A\Gamma^{-1}A^T)^{-1}\hat{a}.$$

Substituting this ν into $\nu^T\Gamma\nu$ gives (2.6.4) and the first conclusion of theorem 2.6.1 holds.

Suppose then that we are in the second case. If there is a ϕ so that $\phi^T\Delta\phi = 0$ but $\nu^T\Gamma A_\perp\phi \neq 0$, then (2.6.9) is not bounded below along the line ξ a multiple of ϕ. We shall show that if such a ϕ does not exist, then (2.6.5) holds. We can write $\nu_0 = A^r\hat{a}$. Suppose then that $\phi^T A_\perp^T\Gamma A_\perp\phi = 0$, which implies that $\nu_0^T\Gamma A_\perp\xi = (A^r\hat{a})^T\Gamma A_\perp\xi = 0$. But then this is (2.6.5).

It remains only to show (2.6.6). But we know that a minimum exists under this assumption. Thus (2.6.12) is consistent and we may get a value of $\lambda = 2(A\Gamma^{-1}A^T)^\dagger\hat{a}$ and (2.6.6) follows. □

Lemma 2.2.3 is a special case of theorem 2.6.1. To see this, let

$$\overline{\nu} = \begin{pmatrix} \nu \\ z \end{pmatrix}, \ \overline{\Gamma} = \begin{pmatrix} I & 0 \\ 0 & -I \end{pmatrix} = J, \ \overline{A} = G_2 = \begin{pmatrix} G_{12} & 0 \\ G_{22} & -I \end{pmatrix}, \quad (2.6.13)$$

$$\overline{a} = G_1\begin{pmatrix} u \\ y \end{pmatrix} = \begin{pmatrix} -G_{11} & I \\ -G_{21} & -I \end{pmatrix}\begin{pmatrix} u \\ y \end{pmatrix}. \quad (2.6.14)$$

Then the optimization problem (2.2.35), (2.2.37) can be written as

$$\min\left\{\overline{\nu}^T\overline{\Gamma}\overline{\nu}\right\},$$
$$\overline{A}\overline{\nu} = \overline{a},$$

which is (2.6.1), (2.6.2). The Δ in theorem 2.6.1 is $G_{2\perp}^T JG_{2\perp}$, which is positive definite by the assumption of lemma 2.2.3. Substituting (2.6.13), (2.6.14) into (2.6.4) gives (2.2.40) of lemma 2.2.3.

In later chapters we shall be interested in constrained problems such as (2.6.1), (2.6.2) where a runs over a set of measured outputs and inputs. If we want the solution to be finite for all possible a, then the situation greatly simplifies.

Corollary 2.6.1 *The optimization problem (2.6.1), (2.6.2) has a finite solution for all a if and only if A has full row rank and either A is invertible or $\Delta > 0$ where $\Delta = D_\perp^T\Gamma D_\perp$.*

Proof. We need only show that (2.6.5) does not occur. Suppose that (2.6.5) does hold for all a. Suppose $\Delta\phi = 0$ for some vector ϕ. Then $(a - Ac)^T(A^r)^T\Gamma A_\perp\phi = 0$. Since $a - Ac$ is arbitrary we have $(A^r)^T\Gamma A_\perp\phi = 0$. Combining these two ϕ equations gives

$$\begin{pmatrix} A_\perp & A^r \end{pmatrix}^T \Gamma A_\perp\phi = 0.$$

But $\begin{pmatrix} A_\perp & A^r \end{pmatrix}$ and Γ are invertible and $A_\perp$ has full column rank so that $\phi = 0$. □

2.6.2 Dynamic Optimization Problem

Consider the following optimization problem

$$J(s, a, b) = \min_{x(0), \nu} (x(0) - x_0)^T P_0^{-1}(x(0) - x_0) + \int_0^s \nu^T \Gamma \nu \, dt \qquad (2.6.15)$$

subject to

$$\dot{x} = Ax + B\nu + a, \qquad (2.6.16)$$

$$b = Cx + D\nu \qquad (2.6.17)$$

over the interval $[0, T]$. The matrices A, B, C, and D have piecewise continuous in time entries. D has full row rank, $P_0 > 0$, and Γ is symmetric and invertible but not necessarily sign definite. The vectors a and b are known continuous functions of time.

Theorem 2.6.2 *The optimization problem (2.6.15) has a bounded unique solution for all $s \in (0, T]$ if and only if*

1. *Either D is invertible or*

$$D_\perp^T \Gamma D_\perp > 0, \quad \forall\, t \in [0, T] \qquad (2.6.18)$$

where $D_\perp$ is a highest-rank matrix such that $DD_\perp = 0$;

2. *The Riccati equation*

$$\dot{P} = (A - SR^{-1}C)P + P(A - SR^{-1}C)^T$$
$$- PC^T R^{-1} CP + Q - SR^{-1}S^T, \quad P(0) = P_0, \qquad (2.6.19)$$

where

$$\begin{pmatrix} Q & S \\ S^T & R \end{pmatrix} = \begin{pmatrix} B \\ D \end{pmatrix} \Gamma^{-1} \begin{pmatrix} B \\ D \end{pmatrix}^T, \qquad (2.6.20)$$

has a solution on $[0, T]$.

Corollary 2.6.2 *If it exists, the unique bounded solution to the optimization problem (2.6.15) is given by*

$$J(s, a, b) = \int_0^s \mu^T R^{-1} \mu \, dt, \qquad (2.6.21)$$

where

$$\mu = C\hat{x} - b, \qquad (2.6.22)$$

and where $\hat{x}$ satisfies

$$\dot{\hat{x}} = A\hat{x} - (S + PC^T)R^{-1}\mu + a, \quad \hat{x}(0) = x_0, \qquad (2.6.23)$$

with P the solution of (2.6.19).

Proof. We use the method of dynamic programming [6] to prove theorem 2.6.2 and corollary 2.6.2. In particular, we use the forward dynamic programming approach by defining the past cost as $W(t,x)$. This is defined exactly as $J(t,a,b)$ but with the additional constraint $x(t) = x$. Then the forward dynamic programming equation becomes

$$\frac{\partial W}{\partial t} = \min_{\nu} -\frac{\partial W}{\partial x}(Ax + B\nu + a) + \nu^T \Gamma \nu, \qquad (2.6.24)$$

where the minimization is subject to (2.6.17). We also have

$$W(0,x) = (x(0) - x_0)^T P_0^{-1}(x(0) - x_0). \qquad (2.6.25)$$

We suppose that $W(t,x)$ can be expressed as follows:

$$W(t,x) = (x - z)^T \Pi^{-1}(x - z) + \int_0^t \sigma^T X \sigma d\tau. \qquad (2.6.26)$$

We should then determine $z(t), \Pi(t), \sigma(t)$, and X. Clearly, $\Pi(0) = P_0$ and $z(0) = x_0$.

The minimization in (2.6.24) can be expressed as follows:

$$\frac{\partial W}{\partial t} = \min_{\nu} \big\{ -(x - z)^T \Pi^{-1}(Ax + B\nu + a)$$
$$-(Ax + B\nu + a)^T \Pi^{-1}(x - z) + \nu^T \Gamma \nu \big\} \qquad (2.6.27)$$

subject to $b = Cx + D\nu$. We can rewrite this problem by completing the square to give

$$\frac{\partial W}{\partial t} = -(x - z)^T \Pi^{-1} B \Gamma^{-1} B^T \Pi^{-1}(x - z) - (x - z)^T \Pi^{-1}(Ax + a)$$
$$-(Ax + a)^T \Pi^{-1}(x - z)$$
$$+ \min_{\nu}(\nu - \Gamma^{-1} B^T \Pi^{-1}(x - z))^T \Gamma (\nu - \Gamma^{-1} B^T \Pi^{-1}(x - z)) \; (2.6.28)$$

subject to

$$b - Cx = D\nu. \qquad (2.6.29)$$

But this problem is in the form (2.6.1),(2.6.2) and thus we use theorem 2.6.1, which tells us that a unique solution exists if and only if (2.6.18) is satisfied and gives an explicit expression for the minimum. This gives us

$$\frac{\partial W}{\partial t} = (b - Cx - S^T \Pi^{-1}(x - z))^T R^{-1}(b - Cx - S^T \Pi^{-1}(x - z))$$
$$-(x - z)^T \Pi^{-1}(Ax + a) - (Ax + a)^T \Pi^{-1}(x - z)$$
$$-(x - z)^T \Pi^{-1} Q \Pi^{-1}(x - z), \qquad (2.6.30)$$

where Q, S, and R are defined in (2.6.20). But

$$\frac{\partial W}{\partial t} = -\dot{z}^T \Pi^{-1}(x - z) - (x - z)^T \Pi^{-1} \dot{\Pi} \Pi^{-1}(x - z)$$
$$-(x - z)^T \Pi^{-1} \dot{z} + \sigma^T X \sigma. \qquad (2.6.31)$$

From (2.6.30) and (2.6.31), by setting $x = z$, we obtain

$$\sigma^T X \sigma = (b - Cz)^T R^{-1}(b - Cz), \qquad (2.6.32)$$

which means that we can take

$$X = R^{-1} \tag{2.6.33}$$

and

$$\sigma = b - Cz. \tag{2.6.34}$$

Again from (2.6.30) and (2.6.31), isolating the quadratic terms in x, we obtain

$$\dot{\Pi} = A\Pi + \Pi A^T + Q - (C\Pi + S^T)^T R^{-1}(C\Pi + S^T). \tag{2.6.35}$$

But this equation is exactly the same as (2.6.19) (with Π in place of P); thus

$$\Pi = P. \tag{2.6.36}$$

This proves theorem 2.6.2.

To prove Corollary 2.6.2, note that from (2.6.30) and (2.6.31), isolating the linear terms in x, and using (2.6.36) to replace Π with P, we obtain

$$-P^{-1}\dot{z} = (C + S^T P^{-1})^T R^{-1}(Cz - b) - P^{-1}(Az + a), \tag{2.6.37}$$

which yields

$$\dot{z} = Az + a - (CP + S^T)^T R^{-1}(Cz - b). \tag{2.6.38}$$

But this equation is identical to (2.6.23). Thus

$$z = \hat{x}. \tag{2.6.39}$$

The only thing that remains to be shown is (2.6.21). From the dynamic programming method, we know that

$$J(s, a, b) = \min_x W(s, x)$$

$$= \min_x (x - \hat{x}(s))^T P^{-1}(s)(x - \hat{x}(s)) + \int_0^s \mu^T R^{-1} \mu \, dt. \tag{2.6.40}$$

But $P^{-1}(s) > 0$ because $P(0)$ is, and P is symmetric and continuous, so the minimum is attained for $x = \hat{x}(s)$, resulting in (2.6.21). □

Note that the existence condition we consider in theorem 2.6.2 has to do with a family of optimization problems and in particular the optimization problem (2.6.15) for all $s \in (0, T]$. It turns out that the existence condition for the single optimization problem (2.6.15), $s = T$, is not the same in general. To see this, consider the following example.

Example 2.6.1 *Consider the linear system*

$$\dot{x} = 0, \tag{2.6.41}$$

$$y = x + \nu \tag{2.6.42}$$

defined over $[0, T]$, and the optimization problem

$$J(T, y) = \min_{x(0), \nu} \left\{ x(0)^2 + \int_0^T \nu^2 \Gamma(t) \, dt \right\} \tag{2.6.43}$$

where

$$\Gamma(t) = \begin{cases} -1 & \text{if } 0 \le t \le \tau \\ +1 & \text{if } \tau < t \end{cases} \tag{2.6.44}$$

with $\tau > 1$.

To find whether or not the optimization problem (2.6.43) has a unique bounded solution, note that $x = x(0)$ is just a constant. So we can express (2.6.43) as follows:

$$J(T, y) = \min_{x(0)} \left\{ x(0)^2 + \int_0^T (y - x(0))^2 \Gamma(t) \, dt \right\}$$

$$= \min_{x(0)} \left\{ x(0)^2 \left(1 + \int_0^T \Gamma(t) \, dt \right) - 2x(0) \int_0^T \Gamma(t) y \, dt + \int_0^T y^2 \Gamma(t) \, dt \right\}. \tag{2.6.45}$$

We can now consider two different cases.

- $T \le \tau$. *In this case we have*

$$J(T, y) = \min_{x(0)} \left\{ (1 - T)x(0)^2 - (\text{linear terms in } x(0) + \text{constant}) \right\}. \tag{2.6.46}$$

  Thus the problem has a unique solution if and only if $T < 1$.

- $T > \tau$. *In this case we have*

$$J(T, y) = \min_{x(0)} \left\{ (1 - 2\tau + T)x(0)^2 - (\text{linear terms in } x(0) + \text{constant}) \right\}. \tag{2.6.47}$$

  So the problem has a unique solution if and only if $T > 2\tau - 1$.

In conclusion, if $T < 1$ or $T > 2\tau - 1$, a unique solution exists; otherwise, it does not. This is consistent with the result in theorem 2.6.2 because the Riccati equation (2.6.19) in this case becomes

$$\dot{p} = -\Gamma(t)p^2, \quad p(0) = 1. \tag{2.6.48}$$

It is easy to verify that

$$p = \frac{1}{1 - t}, \quad t < 1. \tag{2.6.49}$$

Thus according to the theorem, if $T < 1$, $J(s, y)$ has a bounded unique solution for all $s \in (0, T]$. This is of course consistent with our calculations.

The next result is used when working with model uncertainty.

Theorem 2.6.3 *Suppose that there is a bounded unique solution to the optimization problem (2.6.15) for all $s \in (0, T]$ and consider the same problem as (2.6.15) but with the additional constraint*

$$\bar{b} = \bar{C}x + \bar{D}\nu \tag{2.6.50}$$

where the matrix $\begin{pmatrix} D \\ \bar{D} \end{pmatrix}$ has full row rank. Then this new optimization problem has a bounded unique solution for all $s \in (0, T]$. With the additional constraint $x(0) = x_0$, the optimization problem (2.6.15) has a bounded unique solution for all $s \in (0, T]$ if and only if the same conditions are satisfied with $P_0 = 0$ in (2.6.19).

Proof. We shall verify the two conditions of theorem 2.6.2 for this new system. Note that

$$\ker \begin{pmatrix} D \\ \bar{D} \end{pmatrix} \subset \ker D \qquad (2.6.51)$$

so there exists a matrix X such that

$$\begin{pmatrix} D \\ \bar{D} \end{pmatrix}_\perp = D_\perp X. \qquad (2.6.52)$$

Condition (2.6.18) for the new system is

$$\begin{pmatrix} D \\ \bar{D} \end{pmatrix}_\perp^T \Gamma \begin{pmatrix} D \\ \bar{D} \end{pmatrix}_\perp > 0, \quad \forall\, t \in [0, T], \qquad (2.6.53)$$

which becomes

$$X^T D_\perp^T \Gamma D_\perp X > 0, \quad \forall\, t \in [0, T], \qquad (2.6.54)$$

which holds since $D_\perp^T \Gamma D_\perp > 0$ for all $t \in [0, T]$. Thus the first condition of theorem 2.6.2 is satisfied.

For the second condition, we have to show that the Riccati equation (2.6.19) for the new system does not diverge on $[0, T]$. For that, we will show that the solution of this Riccati equation, denoted P_{new}, satisfies

$$P_{\text{new}}(t) \leq P(t) \qquad (2.6.55)$$

on $[0, T]$, where P is the solution of the Riccati equation (2.6.19) for the original system. Note that a, b, and $\bar{b}$ do not appear in the Riccati equation. Assume that a, b, and $\bar{b}$ are zero. Then we have that

$$W(t, x) = x^T P^{-1}(t)x, \qquad (2.6.56a)$$
$$W_{\text{new}}(t, x) = x^T P_{\text{new}}^{-1}(t)x, \qquad (2.6.56b)$$

where W and W_{new} denote, respectively, the past cost given $x(t) = x$ as defined in the dynamic formulation problem used in the proof of theorem 2.6.2. But the new problem is the same optimization problem as the original with an additional constraint. Thus, for all x and t, we must have

$$W(t, x) \leq W_{\text{new}}(t, x). \qquad (2.6.57)$$

But this, thanks to (2.6.56), implies (2.6.55), which is what we needed to show.

The final statement of the theorem is obtained from the first part by using a penalization method. In particular, one can set P_0 in the cost equal to ϵI and look at the limit as ϵ goes to zero. □

2.6.3 Discrete-Time Optimization Problem

Consider the following optimization problem:

$$J(j, a, b) = \min_{x(0),\omega} (x(0) - \hat{x}_0)^T P_0^{-1}(x(0) - \hat{x}_0) + \sum_{k=0}^{j} \omega^T(k)\Gamma\omega(k) \qquad (2.6.58)$$

subject to

$$x(k+1) = Ax(k) + M\omega(k) + a_k, \tag{2.6.59}$$

$$b_k = Cx(k) + N\omega(k) \tag{2.6.60}$$

for $k \in [0, N-1]$. The matrices A, C, M, N may depend on k. We suppose that N has full row rank, $P_0 > 0$, and Γ is symmetric and invertible but not necessarily sign definite. The vectors a_k and b_k are known.

There are two fundamental ways to approach discrete-time optimal control problems. One is through necessary conditions. The other is through dynamic programming, which is based on the minimization principle of Bellman [6]. Depending on whether there are terminal or initial costs, we get the backward version using cost to go or the forward version using past cost. The principle of optimality says that an optimal policy has the property that no matter what the previous (future) decisions (controls) have been (are), the remaining decisions must be optimal with regard to the state resulting from previous decisions. Thus if we are at any point p on an optimal trajectory, the remaining part of the trajectory will be optimal with respect to starting at p.

Theorem 2.6.4 *The optimization problem (2.6.15) has a bounded unique solution (x, ω) for all $j \in [0, N-1]$ if and only if*

1. Either $\begin{pmatrix} A & M \\ C & N \end{pmatrix}$ *is invertible or*

$$\begin{pmatrix} A & M \\ C & N \end{pmatrix}_{\perp}^{T} \begin{pmatrix} P_k^{-1} & 0 \\ 0 & \Gamma \end{pmatrix} \begin{pmatrix} A & M \\ C & N \end{pmatrix}_{\perp} > 0, \ \forall\, k; \tag{2.6.61}$$

2. The Riccati equation

$$P_{k+1} = (AP_kA^T + M\Gamma^{-1}M^T)$$
$$-(AP_kC^T + M\Gamma^{-1}N^T)(CP_kC^T + N\Gamma^{-1}N^T)^{-1}(AP_kC^T + M\Gamma^{-1}N^T)^T \tag{2.6.62}$$

has a positive solution P_k on $[1, N]$.

Corollary 2.6.3 *The solution to the optimization problem (2.6.15) is given by*

$$J(j, a, b) = \sum_{k=0}^{j} \hat{\mu}(k)^T \Delta_k^{-1} \hat{\mu}(k), \tag{2.6.63}$$

where

$$\Delta_k = CP_kC^T + N\Gamma^{-1}N^T \tag{2.6.64}$$

and

$$\hat{\mu}(k) = C\hat{x}(k) - b_k, \tag{2.6.65}$$

and where $\hat{x}$ satisfies the "Kalman filter" equation

$$\hat{x}(k+1) = A\hat{x}(k) + a_k - (AP_kC^T + M\Gamma^{-1}N^T)(CP_kC^T + N\Gamma^{-1}N^T)^{-1}\hat{\mu}(k),$$
$$\hat{x}(0) = \hat{x}_0, \tag{2.6.66}$$

with P_k the solution of the Riccati equation (2.6.62).

Proof. We shall use forward dynamic programming since there is an initial cost. Let the past cost from $x(j) = x_j$ be defined as

$$V_j(x_j) = \min_{x,\omega} \left\{ (x(0) - \hat{x}_0)^T P_0^{-1}(x(0) - \hat{x}_0) + \sum_{k=0}^{j} \omega^T(k)\Gamma\omega(k) \right\} \quad (2.6.67)$$

subject to (2.6.59), (2.6.60), and $x(j) = x_j$. Then the principle of minimality says that V satisfies the dynamic programming equation

$$V_{j+1}(x_{j+1}) = \min_{x_j,\omega(j)} \{V_j(x_j) + \omega^T(j)\Gamma\omega(j)\} \quad (2.6.68)$$

subject to

$$x_{j+1} = Ax_j + M\omega(j) + a_j, \quad (2.6.69)$$
$$b_j = Cx_j + N\omega(j). \quad (2.6.70)$$

Clearly,

$$V_0(x_0) = (x_0 - \hat{x}(0))^T P_0^{-1}(x_0 - \hat{x}(0)). \quad (2.6.71)$$

For V_1, we obtain

$$V_1(x_1) = \min_{x_0,\omega(0)} \{(x_0 - \hat{x}(0))^T P_0^{-1}(x_0 - \hat{x}(0)) + \omega^T(0)\Gamma\omega(0)\}. \quad (2.6.72)$$

This can be rewritten as follows:

$$V_1(x_1) = \min_{x_0,\omega(0)} \left(\begin{pmatrix} x_0 \\ \omega(0) \end{pmatrix} - \begin{pmatrix} \hat{x}(0) \\ 0 \end{pmatrix} \right)^T \begin{pmatrix} P_0^{-1} & 0 \\ 0 & \Gamma \end{pmatrix} \left(\begin{pmatrix} x_0 \\ \omega(0) \end{pmatrix} - \begin{pmatrix} \hat{x}(0) \\ 0 \end{pmatrix} \right)$$
$$(2.6.73)$$

subject to

$$\begin{pmatrix} x_1 - a_0 \\ b_0 \end{pmatrix} = \begin{pmatrix} A & M \\ C & N \end{pmatrix} \begin{pmatrix} x_0 \\ \omega(0) \end{pmatrix}. \quad (2.6.74)$$

The application of theorem 2.6.1 gives that (2.6.4) holds for $k = 0$ and the minimum cost is

$$\begin{pmatrix} x_1 - a_0 - A\hat{x}(0) \\ b_0 - C\hat{x}(0) \end{pmatrix}^T \begin{pmatrix} \Sigma_0 & \Theta_0 \\ \Theta_0^T & \Delta_0 \end{pmatrix}^{-1} \begin{pmatrix} x_1 - a_0 - A\hat{x}(0) \\ b_0 - C\hat{x}(0) \end{pmatrix}, \quad (2.6.75)$$

where Δ_0 is as above, $\Theta_0 = AP_0C^T + M\Gamma^{-1}N^T$, and $\Sigma_0 = AP_0A^T + M\Gamma^{-1}M^T$. Note that

$$P_1 = \Sigma_0 - \Theta_0\Delta_0^{-1}\Theta_0^T. \quad (2.6.76)$$

But a direct calculation shows that

$$\begin{pmatrix} \Sigma_0 & \Theta_0 \\ \Theta_0^T & \Delta_0 \end{pmatrix} = \begin{pmatrix} I & \Theta_0\Delta_0^{-1} \\ 0 & I \end{pmatrix} \begin{pmatrix} \Sigma_0 - \Theta_0\Delta_0^{-1}\Theta_0^T & 0 \\ 0 & \Delta_0 \end{pmatrix} \begin{pmatrix} I & 0 \\ \Delta_0^{-1}\Theta_0^T & I \end{pmatrix}.$$
$$(2.6.77)$$

Thus

$$\begin{pmatrix} \Sigma_0 & \Theta_0 \\ \Theta_0^T & \Delta_0 \end{pmatrix}^{-1} = \begin{pmatrix} I & 0 \\ -\Delta_0^{-1}\Theta_0^T & I \end{pmatrix} \begin{pmatrix} P_1^{-1} & 0 \\ 0 & \Delta_0^{-1} \end{pmatrix} \begin{pmatrix} I & -\Theta_0\Delta_0^{-1} \\ 0 & I \end{pmatrix}. \quad (2.6.78)$$

Combining the terms (2.6.76), (2.6.75), (2.6.78) and using (2.6.65) we get that

$$V_1(x_1) = (x_1 - a_0 - A\hat{x}(0) + \Theta_0\Delta_0^{-1}\hat{\mu}_0)^T P_1^{-1}(x_1 - a_0 - A\hat{x}(0) + \Theta_0\Delta_0^{-1}\hat{\mu}_0)$$
$$+ \mu^T \Delta_0^{-1}\mu_0. \tag{2.6.79}$$

The optimal solution $J(0, a, b)$ is obtained by minimizing this expression over x_1. Thus we have

$$x_1 - a_0 - A\hat{x}(0) + \Theta_0\Delta_0^{-1}\hat{\mu}_0 = 0,$$

which is (2.6.66), and the cost is $\mu^T\Delta_0^{-1}\mu_0$, which is (2.6.63). This establishes the formulas for $j = 0$.

Having determined V_1 we return to (2.6.68) and repeat the process using the same linear algebra as before. In every step one just replaces a subscript 1 by $j + 1$ and a subscript 0 by j. □

2.6.3.1 Descriptor Formulation

The optimization problem considered in section 2.6.3 can be formulated and solved in a more general setting using a descriptor system formulation [15, 29]. Descriptor systems contain mixtures of dynamic and static (or algebraic) equations. We shall give one result concerning descriptor systems and show how it can be used to handle two variants of the robust detection problem for discrete systems.

Consider the following optimization problem:

$$J(j, a) = \min_{x(0),\omega} \left\{ (x(0) - \hat{x}_0)^T P_0^{-1}(x(0) - \hat{x}_0) + \sum_{k=0}^{j} \omega^T(k)\Gamma\omega(k) \right\} \tag{2.6.80a}$$

subject to

$$Ex(k + 1) = Fx(k) + G\omega(k) + a_k \tag{2.6.80b}$$

for $k \in [0, N - 1]$. The matrices E, F, G may depend on k and E is not assumed to be square or invertible. We suppose that $(F \ \ G)$ has full row rank, E has full column rank, $P_0 > 0$, and Γ is symmetric and invertible but not necessarily sign definite. The vectors a_k are known.

Theorem 2.6.5 *The optimization problem (2.6.80) has a unique bounded solution* (x, ω) *for all* $j \in [0, N - 1]$ *if and only if*

1. *Either* $(F \ \ G)$ *is invertible or*

$$(F \ \ G)_\perp^T \begin{pmatrix} P_k^{-1} & 0 \\ 0 & \Gamma \end{pmatrix} (F \ \ G)_\perp > 0, \ \forall\, k; \tag{2.6.81}$$

2. *The Riccati equation*

$$Q_k = FP_kF^T + G\Gamma^{-1}G^T, \tag{2.6.82}$$
$$P_{k+1} = E^TQ_k^{-1}E, \ P(0) = P_0 \tag{2.6.83}$$

has positive solution P_k *on* $[1, N]$.

Corollary 2.6.4 *The solution to the optimization problem (2.6.80a) is given by*

$$J(j,a) = \sum_{k=0}^{j} \hat{\mu}(k)^T \Delta_k \hat{\mu}(k) \tag{2.6.84}$$

where

$$\Delta_k = Q_k^{-1} - Q_k^{-1} E (E^T Q_k^{-1} E)^{-1} E^T Q_k^{-1} \tag{2.6.85}$$

and

$$\hat{\mu}(k) = a_k + F\hat{x}_k, \tag{2.6.86}$$

and where $\hat{x}$ satisfies the "Kalman filter" equation

$$\hat{x}(k+1) = (E^T Q_k^{-1} E)^{-1} E^T Q_k^{-1} \hat{\mu}(k), \quad \hat{x}(0) = \hat{x}_0, \tag{2.6.87}$$

with P_k the solution of (2.6.83).

Proof. We again use forward dynamic programming. Let the past cost from $x(j) = x_j$ be defined as

$$V_j(x_j) = \min_{x,\omega} \left\{ (x(0) - \hat{x}_0)^T P_0^{-1}(x(0) - \hat{x}_0) + \sum_{k=0}^{j} \omega^T(k) \Gamma \omega(k) \right\} \tag{2.6.88}$$

subject to (2.6.80b) and $x(j) = x_j$. Then the principle of minimality says that V satisfies the dynamic programming equation

$$V_{j+1}(x_{j+1}) = \min_{x_j, \omega(j)} \left\{ V_j(x_j) + \omega^T(j) \Gamma \omega(j) \right\} \tag{2.6.89}$$

subject to

$$E x_{j+1} - a_j = F x_j + G\omega(j). \tag{2.6.90}$$

Clearly,

$$V_0(x_0) = (x_0 - \hat{x}(0))^T P_0^{-1}(x_0 - \hat{x}(0)). \tag{2.6.91}$$

The result then follows from the recursive application of theorem 2.6.1. □

Corollary 2.6.5 *Suppose that E is not of full row rank. Let $E^\perp$ be a maximal rank full row rank matrix such that $E^\perp E = 0$. Then*

$$J(j,a) = \sum_{k=0}^{j} \sigma(k)^T \Omega_k \sigma(k), \tag{2.6.92}$$

$$\sigma(k) = E^\perp(a_k + F\hat{x}_k), \tag{2.6.93}$$

and

$$\Omega_k = (E^\perp Q_k E^{\perp T})^{-1}. \tag{2.6.94}$$

Proof. What we need to show here is that

$$E^{\perp T}(E^\perp Q E^{\perp T})^{-1} E^\perp = Q^{-1} - Q^{-1} E (E^T Q^{-1} E)^{-1} E^T Q^{-1}. \qquad (2.6.95)$$

By multiplying both sides of (2.6.95) with

$$X = \begin{pmatrix} E & Q E^{\perp T} \end{pmatrix} \qquad (2.6.96)$$

on the right, we obtain

$$\begin{pmatrix} 0 & E^{\perp T} \end{pmatrix} = \begin{pmatrix} 0 & E^{\perp T} \end{pmatrix}. \qquad (2.6.97)$$

Thus all we need to show is that X is invertible. Consider

$$\begin{pmatrix} E^T \\ E^\perp \end{pmatrix} X = \begin{pmatrix} E^T \\ E^\perp \end{pmatrix} \begin{pmatrix} E & Q E^{\perp T} \end{pmatrix} = \begin{pmatrix} E^T Q^{-1} E & 0 \\ E^\perp Q^{-1} E & E^\perp E^{\perp T} \end{pmatrix}, \qquad (2.6.98)$$

which clearly is invertible. Thus X is invertible.   □

We now show how these results can be used for the robust detection problem of section 2.4. We consider the following two situations:

$$x(k+1) = Ax(k) + Bu(k) + M\nu(k), \qquad (2.6.99)$$
$$y(k) = Cx(k) + Du(k) + N\nu(k), \qquad (2.6.100)$$
$$z(k) = Kx(k) + Hu(k) + J\nu(k), \qquad (2.6.101)$$

which we call the nondelayed system, and

$$x(k+1) = Ax(k) + Bu(k) + M\nu(k), \qquad (2.6.102)$$
$$y(k+1) = Cx(k) + Du(k) + N\nu(k), \qquad (2.6.103)$$
$$z(k) = Kx(k) + Hu(k) + J\nu(k), \qquad (2.6.104)$$

which we call the delayed system because of the delay between the state and the measured output. Such delays are common in practice. In both cases, we consider the following optimization problem:

$$J(j, u, y) = \min_{x(0), \nu} \left\{ (x(0) - \hat{x}_0)^T P_0^{-1} (x(0) - \hat{x}_0) + \sum_{k=0}^{j} |\nu(k)|^2 - |z(k)|^2 \right\}. \qquad (2.6.105)$$

Theorem 2.6.5 can be used directly to solve both problems.

System without Delay

In this case, we simply let

$$\nu = \begin{pmatrix} \nu \\ z \end{pmatrix}, \ \Gamma = \begin{pmatrix} I & 0 \\ 0 & -I \end{pmatrix}, \ a_k = \begin{pmatrix} B & 0 \\ D & -I \\ H & 0 \end{pmatrix} \begin{pmatrix} u(k) \\ y(k) \end{pmatrix}, \qquad (2.6.106)$$

and

$$E = \begin{pmatrix} I \\ 0 \\ 0 \end{pmatrix}, \ F = \begin{pmatrix} A \\ C \\ K \end{pmatrix}, \ G = \begin{pmatrix} M & 0 \\ N & 0 \\ J & -I \end{pmatrix}. \qquad (2.6.107)$$

Then the positivity test (2.6.81) becomes

$$
\begin{pmatrix} A & M & 0 \\ C & N & 0 \\ K & J & -I \end{pmatrix}^T_\perp \begin{pmatrix} P_k^{-1} & 0 & 0 \\ 0 & I & 0 \\ 0 & 0 & -I \end{pmatrix} \begin{pmatrix} A & M & 0 \\ C & N & 0 \\ K & J & -I \end{pmatrix}_\perp > 0, \quad \forall\, k. \quad (2.6.108)
$$

The Riccati equation in this case becomes

$$
P_{k+1} = \begin{pmatrix} I \\ 0 \\ 0 \end{pmatrix}^T Q_k^{-1} \begin{pmatrix} I \\ 0 \\ 0 \end{pmatrix}, \qquad (2.6.109)
$$

where

$$
Q_k = \begin{pmatrix} A & M & 0 \\ C & N & 0 \\ K & J & -I \end{pmatrix} \begin{pmatrix} P_k^{-1} & 0 & 0 \\ 0 & I & 0 \\ 0 & 0 & -I \end{pmatrix} \begin{pmatrix} A & M & 0 \\ C & N & 0 \\ K & J & -I \end{pmatrix}^T. \qquad (2.6.110)
$$

The estimate of x can be computed as follows:

$$
\hat{x}(k+1) = P_{k+1}^{-1} \begin{pmatrix} I \\ 0 \\ 0 \end{pmatrix}^T Q_k^{-1} \begin{pmatrix} A\hat{x}(k) + Bu(k) \\ C\hat{x}(k) + Du(k) - y(k) \\ K\hat{x}(k) + Hu(k) \end{pmatrix}. \qquad (2.6.111)
$$

In this case, we can let

$$
E^\perp = \begin{pmatrix} 0 & I & 0 \\ 0 & 0 & I \end{pmatrix} \qquad (2.6.112)
$$

which gives in (2.6.92)

$$
\sigma(k) = \begin{pmatrix} C\hat{x}(k) + Du(k) - y(k) \\ K\hat{x}(k) + Hu(k) \end{pmatrix} \qquad (2.6.113)
$$

and

$$
\Omega_k = \left(\begin{pmatrix} 0 & I & 0 \\ 0 & 0 & I \end{pmatrix} Q_k \begin{pmatrix} 0 & I & 0 \\ 0 & 0 & I \end{pmatrix}^T \right)^{-1}. \qquad (2.6.114)
$$

Straightforward algebra shows that this result is consistent with theorem 2.4.2 and its corollary.

Delayed System

In this case, we also let

$$
\nu = \begin{pmatrix} \nu \\ z \end{pmatrix}, \; \Gamma = \begin{pmatrix} I & 0 \\ 0 & -I \end{pmatrix}. \qquad (2.6.115)
$$

But here

$$
a_k = \begin{pmatrix} B & 0 \\ D & -I \\ H & 0 \end{pmatrix} \begin{pmatrix} u(k) \\ y(k+1) \end{pmatrix}, \qquad (2.6.116)
$$

and

$$E = \begin{pmatrix} I \\ -C \\ 0 \end{pmatrix}, \; F = \begin{pmatrix} A \\ 0 \\ K \end{pmatrix}, \; G = \begin{pmatrix} M & 0 \\ N & 0 \\ J & -I \end{pmatrix}. \tag{2.6.117}$$

So the positivity test (2.4.30) becomes

$$\begin{pmatrix} A & M & 0 \\ 0 & N & 0 \\ K & J & -I \end{pmatrix}_{\perp}^{T} \begin{pmatrix} P_k^{-1} & 0 & 0 \\ 0 & I & 0 \\ 0 & 0 & -I \end{pmatrix} \begin{pmatrix} A & M & 0 \\ 0 & N & 0 \\ K & J & -I \end{pmatrix}_{\perp} > 0, \; \forall k \tag{2.6.118}$$

and the Riccati equation in this case becomes

$$P_{k+1} = \begin{pmatrix} I \\ -C \\ 0 \end{pmatrix}^{T} Q_k^{-1} \begin{pmatrix} I \\ -C \\ 0 \end{pmatrix}, \tag{2.6.119}$$

where

$$Q_k = \begin{pmatrix} A & M & 0 \\ 0 & N & 0 \\ K & J & -I \end{pmatrix} \begin{pmatrix} P_k^{-1} & 0 & 0 \\ 0 & I & 0 \\ 0 & 0 & -I \end{pmatrix} \begin{pmatrix} A & M & 0 \\ 0 & N & 0 \\ K & J & -I \end{pmatrix}^{T}. \tag{2.6.120}$$

The estimate of x can be computed as follows:

$$\hat{x}(k+1) = P_{k+1}^{-1} \begin{pmatrix} I \\ -C \\ 0 \end{pmatrix}^{T} Q_k^{-1} \begin{pmatrix} A\hat{x}(k) + Bu(k) \\ Du(k) - y(k+1) \\ K\hat{x}(k) + Hu(k) \end{pmatrix}. \tag{2.6.121}$$

In this case, we can let

$$E^{\perp} = \begin{pmatrix} C & I & 0 \\ 0 & 0 & I \end{pmatrix} \tag{2.6.122}$$

which gives in (2.6.92)

$$\sigma(k) = \begin{pmatrix} C A\hat{x}(k) + (CB + D)u(k) - y(k+1) \\ K\hat{x}(k) + Hu(k) \end{pmatrix} \tag{2.6.123}$$

and

$$\Omega_k = \left(\begin{pmatrix} C & I & 0 \\ 0 & 0 & I \end{pmatrix} Q_k \begin{pmatrix} C & I & 0 \\ 0 & 0 & I \end{pmatrix}^{T} \right)^{-1}. \tag{2.6.124}$$

Chapter Three

Multimodel Formulation

3.1 INTRODUCTION

By modeling the normal and the failed behaviors of a process using two or more linear uncertain systems, failure detectability in linear systems can be seen as a linear multimodel identification problem. In most cases, there is no guarantee that one of the models can be ruled out by simply observing the inputs and outputs of the system. For example, if the input to a linear system at rest is zero, the output remains zero, and zero input/output is consistent with both models. We shall also see examples later in chapter 4 where the failure is not apparent during normal operations. For example, a failure of the brakes may not be apparent while driving unless the driver performs a special test, such as tapping the brakes lightly.

It is for this reason that in some cases a test signal, usually referred to as an *auxiliary signal*, is injected into the system to expose its behavior and facilitate the detection (identification) of the failure. This means that the inputs of the system are partially or completely taken over by the failure detector mechanism for the period of time during which the auxiliary signal is injected into the system, and the failure detection test is performed based on measurements of the inputs and outputs.

In this chapter we shall focus on the case where there are two possible models, which we shall call model 0 and model 1. We let v represent the inputs taken over by the failure detector mechanism, u the rest of the inputs, if any, and y the outputs of the system. Then an auxiliary signal v guarantees failure detection if and only if

$$\mathcal{A}_0(v) \cap \mathcal{A}_1(v) = \emptyset \qquad (3.1.1)$$

where $\mathcal{A}_i(v)$ represents the set of input-output pairs $\{u, y\}$ consistent with model i, $i = 0, 1$, for a given input v. We call such a v a *proper auxiliary signal*.

The approach of this chapter can be applied to problems with more than one type of failure by designing a test signal for each pair of models and applying the test signals sequentially. In chapter 4 we show how to design one signal to test simultaneously for several different types of failure.

If an auxiliary signal satisfying (3.1.1) exists, it is usually not difficult to find one, since unreasonably "large" signals often do the job. But such signals cannot be applied in practice. For example, if the auxiliary signal is a mechanical force, there are always limitations on it (actuator characteristics). Similarly, if it is an electrical signal, there are limitations imposed by the electrical components used. But there are even more stringent conditions in practice imposed by the desire that the system, during the test period, should continue to operate in a reasonable manner. For that requirement to be met the test period should be short and the effect of the auxiliary signal on the system minimal. For example, in many cases it is required that the test signal have minimum energy, but more general criteria are also important. Some of these more general criteria will be considered in section 3.5, which discusses more general cost functions.

In this chapter we develop a method for constructing optimal auxiliary signals in the two-model case. We start by considering the static case first to illustrate the fundamental ideas in a simple setting where it is easier to visualize geometrically what is happening. We then consider dynamical systems over a finite test period, as we did in the previous chapter.

3.2 STATIC CASE

In this section we consider a model similar to that of (2.2.52),(2.2.53) in the previous chapter. However, we generalize it slightly as follows:

$$
(X \quad Y \quad Z) \begin{pmatrix} v \\ u \\ y \end{pmatrix} = H\nu \tag{3.2.1a}
$$

with the constraint

$$
\nu^T J \nu < d. \tag{3.2.1b}
$$

Here J is a signature matrix. That is, J is a diagonal matrix with $+1$ and -1 entries. The number of $+1$ and -1 entries is arbitrary so that the two extreme cases are $J = I$ and $J = -I$. However, we shall see later that some additional restrictions on J are necessary in order for a proper auxiliary signal to exist. For example, $J_i = -I$ means there is no restriction at all on the noise and if H_i is full row rank the $\mathcal{A}_i(v)$ will be the whole space. The vectors u and y are input and output as before, but we have included an additional input vector v which represents the auxiliary signal. X, Y, Z, and H are matrices with appropriate dimensions.

The model (3.2.1) can be obtained from the model (2.2.52),(2.2.53) by combining ν and z into a single vector ν and breaking up the input u into two parts, u and v. This model is slightly generalized from that of chapter 2 because we do not impose the form (2.2.36) on the signature matrix J. Thus it is also a generalization of (2.2.4),(2.2.5). This formulation includes both additive and model uncertainty.

We also note that assuming this structure on J is not restrictive. In several of our applications it is already in this form. But if instead we have that $\bar{\nu}^T K \bar{\nu}$ is any quadratic form in $\bar{\nu}$, then we may assume K is symmetric. Then there exists an invertible matrix R such that $R^T K R = J$ and letting $\bar{\nu} = R\nu$ gives (3.2.1b). The H matrix in (3.2.1a) is then HR^{-1}.

We suppose that the failed and unfailed systems can both be represented by models in the form (3.2.1). Thus we have

$$
(X_i \quad Y_i \quad Z_i) \begin{pmatrix} v \\ u \\ y \end{pmatrix} = H_i \nu_i \tag{3.2.2a}
$$

with the constraint

$$\nu_i^T J_i \nu_i < 1, \tag{3.2.2b}$$

for $i = 0, 1$. The case $i = 0$ corresponds to the unfailed system and the case $i = 1$ to the failed system. Note that each system has its own noise vector ν_i of possibly different size, but everything that can be observed by the detector (v, u, and y) is common to both systems. The H_i are assumed to have full row rank.

Suppose we have access to u and y, given a v consistent with one of the models. The problem we consider here is to find an optimal v in some sense for which observation of u and y provides enough information to decide from which model they have been generated. That is, the condition (3.1.1) is satisfied.

If the optimization is on the norm of v, our problem becomes that of finding a v of smallest norm for which there exist no solutions to (3.2.2) for $i = 0$ and 1 simultaneously.

Example 3.2.1 *Consider the following two simple systems:*

$$-4v + y = \nu_0, \tag{3.2.3a}$$
$$\nu_0^2 < 1 \tag{3.2.3b}$$

and

$$-\begin{pmatrix} 1 \\ 1 \end{pmatrix} v + \begin{pmatrix} 1 \\ 0 \end{pmatrix} y = \begin{pmatrix} 1 & 0 \\ 0 & -1 \end{pmatrix} \nu_1, \tag{3.2.4a}$$

$$\nu_1^T \begin{pmatrix} 1 & 0 \\ 0 & -1 \end{pmatrix} \nu_1 < 1. \tag{3.2.4b}$$

Here u is absent. The corresponding realizable sets are given in figure 3.2.1. For a given auxiliary signal $\hat{v}$, the output set $\mathcal{A}_i(\hat{v})$ is given by the intersection of the line $v = \hat{v}$ with the realizable set. A least norm v for which the realizable y's corresponding to the two models do not have any common point is labeled v^*. By symmetry, $-v^*$ is another minimum.

We turn now to developing a more easily computable characterization of proper auxiliary signals. Since the H_i's are full row rank, for any v, u, and y, there always exist ν_i satisfying (3.2.2a). Thus the nonexistence of a solution to (3.2.2a) and (3.2.2b) is equivalent to the noise that is required by one of the models to produce a y consistent with both models being always too large. That is,

$$\sigma(v) \geq 1 \tag{3.2.5}$$

where

$$\sigma(v) = \inf_{\nu_0, \nu_1, u, y} \max(\nu_0^T J_0 \nu_0, \nu_1^T J_1 \nu_1) \tag{3.2.6}$$

subject to (3.2.2a), $i = 0, 1$.

Throughout this chapter we shall make frequent use of the fact that the maximum of two numbers $\{\eta_1, \eta_2\}$ can be rewritten as

$$\max\{\eta_1, \eta_2\} = \max_{0 \leq \beta \leq 1} \{\beta\eta_1 + (1 - \beta)\eta_2\}. \tag{3.2.7}$$

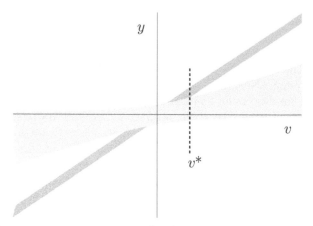

Figure 3.2.1 The realizable sets $\{v, y\}$ corresponding to two models.

Using (3.2.7), we can rewrite (3.2.6) as

$$\sigma(v) = \inf_{\nu_0,\nu_1,u,y} \max_{\beta \in [0,1]} (\beta \nu_0^T J_0 \nu_0 + (1-\beta) \nu_1^T J_1 \nu_1). \qquad (3.2.8)$$

The key to the algorithms that follow is to be able to rewrite the "inf max" of (3.2.8) as a "max inf." The next lemma shows that this interchange is possible.

Lemma 3.2.1 *The function $\sigma(v)$ in (3.2.6) can be expressed as follows:*

$$\sigma(v) = \max_{\beta \in [0,1]} \min_{\nu_0,\nu_1,u,y} (\beta \nu_0^T J_0 \nu_0 + (1-\beta) \nu_1^T J_1 \nu_1) \qquad (3.2.9a)$$

subject to

$$\begin{pmatrix} X_0 & Y_0 & Z_0 \\ X_1 & Y_1 & Z_1 \end{pmatrix} \begin{pmatrix} v \\ u \\ y \end{pmatrix} = \begin{pmatrix} H_0 & 0 \\ 0 & H_1 \end{pmatrix} \begin{pmatrix} \nu_0 \\ \nu_1 \end{pmatrix}. \qquad (3.2.9b)$$

Proof. Since H_0 and H_1 have full row rank, there exist ν_0, ν_1 satisfying (3.2.9b) for any given u, y, v. Equation (3.2.9b) is linear. Thus all the ν_i's corresponding to a given v can be expressed as

$$\nu_i = F_i x + G_i v, \quad i = 0, 1, \qquad (3.2.10)$$

for given matrices F_i, G_i and arbitrary x where

$$F_i x = \begin{pmatrix} F_{i1} & | & F_{i2} \end{pmatrix} \begin{pmatrix} u \\ \hline y \\ \hline z \end{pmatrix}, \quad F_{i2} = \begin{pmatrix} H_0 & 0 \\ 0 & H_1 \end{pmatrix}_{\perp}.$$

Substituting (3.2.10) for the ν_i's in (3.2.6), we see that the optimization problem (3.2.6) is the inf max of quadratic functions. Lemma 3.2.1 then follows from the results of section 3.9.1. $\qquad \square$

Let

$$\phi_\beta(v) = \min_{\nu_0,\nu_1,u,y} (\beta \nu_0^T J_0 \nu_0 + (1-\beta) \nu_1^T J_1 \nu_1). \qquad (3.2.11)$$

Then the optimization problem (3.2.6) can now be written

$$\sigma(v) = \max_{0 \le \beta \le 1} \phi_\beta(v). \tag{3.2.12}$$

In order for there to be a proper v we need that $\sigma(v) \ge 1$. From theorem 2.6.1 we have that $\phi_\beta(v)$ is quadratic in v. That is, $\phi_\beta(cv) = c^2 \phi_\beta(v)$ for scalars c. Thus there will exist a proper v if and only if there are a v and a β such that $\phi_\beta(v) > -\infty$. We now examine when this occurs.

Since u and y appear only in the constraint (3.2.9b), we can eliminate them from the optimization problem by premultiplying (3.2.9b) with

$$\begin{pmatrix} W_0 & W_1 \end{pmatrix} = \begin{pmatrix} Y_0 & Z_0 \\ Y_1 & Z_1 \end{pmatrix}^\perp. \tag{3.2.13}$$

The optimization problem (3.2.11) then becomes

$$\phi_\beta(v) = \inf_\nu \nu^T J_\beta \nu \tag{3.2.14a}$$

subject to

$$Gv = H\nu, \tag{3.2.14b}$$

where

$$G = \begin{pmatrix} W_0 & W_1 \end{pmatrix} \begin{pmatrix} X_0 \\ X_1 \end{pmatrix}, \quad H = \begin{pmatrix} W_0 & W_1 \end{pmatrix} \begin{pmatrix} H_0 & 0 \\ 0 & H_1 \end{pmatrix}, \tag{3.2.15}$$

$$J_\beta = \begin{pmatrix} \beta J_0 & 0 \\ 0 & (1 - \beta) J_1 \end{pmatrix}, \quad \nu = \begin{pmatrix} \nu_0 \\ \nu_1 \end{pmatrix}. \tag{3.2.16}$$

But problem (3.2.14) is solved in theorem 2.6.1.

3.2.1 Proper Auxiliary Signal and the Separability Index

In the previous section, we saw that an auxiliary signal v separates the two output sets $\mathcal{A}_0(v)$ and $\mathcal{A}_1(v)$, that is, it allows for perfect detection, if and only if $\sigma(v) \ge 1$. We thus obtain the following characterization of a proper auxiliary signal.

Lemma 3.2.2 *An auxiliary signal v is proper if and only if $\sigma(v) \ge 1$.*

We want to determine the minimum proper auxiliary signal. In general, in this book we determine minimality with respect to an inner product norm of v. Every inner product norm on R^n may be written as $(v^T Q v)^{1/2}$ for a positive definite matrix Q. Thus we may consider minimizing $v^T Q v$ subject to the constraint that $\sigma(v) \ge 1$. In discussing this optimization problem it will be convenient to define the following quantity.

Definition 3.2.1 *Let $\mathcal{V}$ denote the set of proper auxiliary signals v. Then*

$$\gamma^* = \left(\min_{v \in \mathcal{V}} v^T Q v \right)^{-1/2} \tag{3.2.17}$$

is called the separability index for positive definite matrix Q. The v realizing the miminization in (3.2.17) is called an optimal proper auxiliary signal. An optimal v exists if $\mathcal{V}$ is not empty.

Using the fact that $\sigma(v)$ is quadratic in v, the problem of finding an optimal auxiliary signal can be reformulated into a more computationally useful form as follows. Define the quantity λ^* by

$$\lambda^* = \max_{v \neq 0} \frac{\sigma(v)}{v^T Q v} = \max_{v \neq 0, \beta \in [0,1]} \frac{\phi_\beta(v)}{v^T Q v} = \max_{v \neq 0, \beta \in [0,1]} \frac{v^T V_\beta v}{v^T Q v} \tag{3.2.18}$$

for some symmetric matrix V_β. The second equality in (3.2.17) follows from equation (3.2.12). The third equality follows from $\phi_\beta(v)$ being quadratic in v.

The maximum in (3.2.18) is either $-\infty$ or a finite number. Assuming that $\lambda^* > -\infty$ we may restrict ourselves to considering β where (see theorem 2.6.1)

$$H_\perp^T J_\beta H_\perp > 0. \tag{3.2.19}$$

Let

$$\lambda_\beta = \max_{v \neq 0} \frac{v^T V_\beta v}{v^T Q v}. \tag{3.2.20}$$

Then clearly $\lambda^* = \max_\beta \lambda_\beta$.

Lemma 3.2.3 *Let λ_β be defined by (3.2.20) and assume (3.2.19). Then*

$$\lambda_\beta = \text{largest eigenvalue of } Q^{-1} V_\beta. \tag{3.2.21}$$

In the special case that V_β is positive definite, we have (3.2.21) is the spectral radius, that is,

$$\lambda_\beta = \rho(Q^{-1} V_\beta). \tag{3.2.22}$$

In addition, the v's that give the optimum satisfy

$$(V_\beta - \lambda Q)v = 0 \tag{3.2.23}$$

with $\lambda = \lambda_\beta$.

Proof. Let $z = Q^{1/2} v$, $\widetilde{V}_\beta = Q^{-1/2} V_\beta Q^{-1/2}$. Then (3.2.20) becomes

$$\lambda_\beta = \max_{z \neq 0} \frac{z^T \widetilde{V}_\beta z}{z^T z}. \tag{3.2.24}$$

But $\widetilde{V}_\beta$ is symmetric. Thus λ_β is the largest eigenvalue of $\widetilde{V}_\beta$ and λ_β is the value of the fraction on the right hand side of (3.2.24) if z is an eigenvector of $\widetilde{V}_\beta$ associated with the eigenvalue λ_β. Thus λ_β is the largest λ such that

$$(\widetilde{V}_\beta - \lambda I)z = 0 \tag{3.2.25}$$

for a nonzero z. Letting $z = Q^{1/2} v$ and multiplying (3.2.25) by $Q^{1/2}$, we get that λ_β is the largest λ such that (3.2.23) holds. Multiplying (3.2.23) by Q^{-1} proves the rest of the lemma. $\square$

Theorem 3.2.1 *If $\lambda^* > 0$, then the set of proper auxiliary signals is not empty and any optimal proper auxiliary signal v is a solution of (3.2.23) with $\lambda = \lambda^*$. In addition, v satisfies*

$$v^T Q v = \frac{1}{\lambda^*}. \tag{3.2.26}$$

If $\lambda^* \leq 0$, then the set of proper auxiliary signals is empty.

Corollary 3.2.1 *If* $\lambda^* > 0$, *then the separability index is given by*

$$\gamma^* = \sqrt{\lambda^*}. \tag{3.2.27}$$

Example 3.2.2 *Consider the problem defined in example 3.2.1. The corresponding* $\phi_\beta(v)$ *can be formulated as follows:*

$$\phi_\beta(v) = \inf_{\nu_0, \nu_{10}, \nu_{11}} \begin{pmatrix} \nu_0 \\ \nu_{10} \\ \nu_{11} \end{pmatrix}^T \begin{pmatrix} \beta & & \\ & 1-\beta & \\ & & -(1-\beta) \end{pmatrix} \begin{pmatrix} \nu_0 \\ \nu_{10} \\ \nu_{11} \end{pmatrix} \tag{3.2.28a}$$

subject to

$$\begin{pmatrix} -4 & 1 \\ -1 & 1 \\ -1 & 0 \end{pmatrix} \begin{pmatrix} v \\ y \end{pmatrix} = \begin{pmatrix} 1 & 0 & 0 \\ 0 & 1 & 0 \\ 0 & 0 & -1 \end{pmatrix} \begin{pmatrix} \nu_0 \\ \nu_{10} \\ \nu_{11} \end{pmatrix}. \tag{3.2.28b}$$

The first step is to find a matrix of maximum rank that removes y *from the constraint. We compute* W *as*

$$W = \begin{pmatrix} 1 \\ 1 \\ 0 \end{pmatrix}^{\perp} = \begin{pmatrix} 0 & 0 & 1 \\ -1 & 1 & 0 \end{pmatrix}. \tag{3.2.29}$$

Premultiplying the constraint (3.2.28b) by W *yields*

$$\begin{pmatrix} -1 \\ 3 \end{pmatrix} v = \begin{pmatrix} 0 & 0 & -1 \\ -1 & 1 & 0 \end{pmatrix} \begin{pmatrix} \nu_0 \\ \nu_{10} \\ \nu_{11} \end{pmatrix}. \tag{3.2.30}$$

The solution to the optimization problem (3.2.28a) can now be obtained using theorem 2.6.1. First note that

$$\Delta = \begin{pmatrix} 0 & 0 & -1 \\ -1 & 1 & 0 \end{pmatrix}_{\perp}^T \begin{pmatrix} \beta & & \\ & 1-\beta & \\ & & -(1-\beta) \end{pmatrix} \begin{pmatrix} 0 & 0 & -1 \\ -1 & 1 & 0 \end{pmatrix}_{\perp}$$

$$= \begin{pmatrix} 1 & 1 & 0 \end{pmatrix} \begin{pmatrix} \beta & & \\ & 1-\beta & \\ & & -(1-\beta) \end{pmatrix} \begin{pmatrix} 1 \\ 1 \\ 0 \end{pmatrix} = 1, \tag{3.2.31}$$

so $\phi_\beta(v)$ *is defined for all* $\beta \in [0,1]$ *and is given by*

$$\phi_\beta(v) = \begin{pmatrix} -1 \\ 3 \end{pmatrix}^T \left(\begin{pmatrix} 0 & -1 \\ 0 & 1 \\ -1 & 0 \end{pmatrix}^T \begin{pmatrix} \beta & & \\ & 1-\beta & \\ & & \beta-1 \end{pmatrix}^{-1} \begin{pmatrix} 0 & -1 \\ 0 & 1 \\ -1 & 0 \end{pmatrix} \right)^{-1} \begin{pmatrix} -1 \\ 3 \end{pmatrix} v^2$$

$$= (9\beta - 1)(1 - \beta)v^2. \tag{3.2.32}$$

Note that $\phi_\beta(v)$ *is quadratic in* v. *The graph of* $\phi_\beta(1)$ *is given in figure 3.2.2.*

Thus $\sigma(v) = \max_{\beta \in [0,1]} \phi_\beta(v)$ *can easily be computed in this case by setting the derivative of* $\phi_\beta(v)$ *with respect to* β *to zero, which yields*

$$-18\beta + 10 = 0 \implies \beta^* = 5/9. \tag{3.2.33}$$

This gives

$$\sigma(v) = \frac{16}{9}v^2 \tag{3.2.34}$$

and thus $v^* = 3/4$.

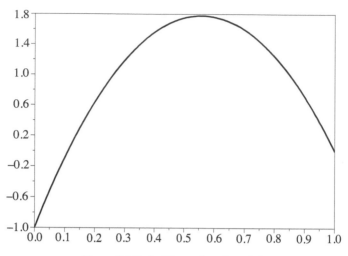

Figure 3.2.2 $\phi_\beta(1)$ as a function of β.

In the above example, the set of β's for which $\phi_\beta(v)$ was finite included the entire interval $[0, 1]$. This is not always the case, as can be seen from the following example.

Example 3.2.3 *Consider the following two systems:*

$$y - u = \nu_{01}, \tag{3.2.35a}$$

$$u = \nu_{02}, \tag{3.2.35b}$$

$$\nu_{01}^2 - \nu_{02}^2 < 1 \tag{3.2.35c}$$

and

$$u = \nu_{11}, \tag{3.2.36a}$$

$$y - v = \nu_{12}, \tag{3.2.36b}$$

$$\nu_{11}^2 + \nu_{12}^2 < 1. \tag{3.2.36c}$$

Putting the two systems together, we obtain

$$\begin{pmatrix} 0 & -1 & 1 \\ 0 & 1 & 0 \\ 0 & 1 & 0 \\ -1 & 0 & 1 \end{pmatrix} \begin{pmatrix} v \\ u \\ y \end{pmatrix} = \begin{pmatrix} \nu_{01} \\ \nu_{02} \\ \nu_{11} \\ \nu_{12} \end{pmatrix}. \tag{3.2.37}$$

Eliminating u and y by multiplying by a left annihilator of the u, y coefficient matrix gives

$$\begin{pmatrix} 0 \\ 1 \end{pmatrix} v = \begin{pmatrix} 0 & 1 & -1 & 0 \\ 1 & 1 & 0 & -1 \end{pmatrix} \begin{pmatrix} \nu_{01} \\ \nu_{02} \\ \nu_{11} \\ \nu_{12} \end{pmatrix}. \tag{3.2.38}$$

In this case, we have

$$J_\beta = \mathrm{diag}(\beta, -\beta, 1 - \beta, 1 - \beta).$$ (3.2.39)

Applying theorem 2.6.1, we find that

$$\Delta = \begin{pmatrix} 0 & 1 & -1 & 0 \\ 1 & 1 & 0 & -1 \end{pmatrix}^T_\perp \begin{pmatrix} \beta & & & \\ & -\beta & & \\ & & 1-\beta & \\ & & & 1-\beta \end{pmatrix} \begin{pmatrix} 0 & 1 & -1 & 0 \\ 1 & 1 & 0 & -1 \end{pmatrix}_\perp$$

$$= \begin{pmatrix} 0 & -1 \\ 1 & 1 \\ 1 & 1 \\ 1 & 0 \end{pmatrix}^T \begin{pmatrix} \beta & & & \\ & -\beta & & \\ & & 1-\beta & \\ & & & 1-\beta \end{pmatrix} \begin{pmatrix} 0 & -1 \\ 1 & 1 \\ 1 & 1 \\ 1 & 0 \end{pmatrix} = \begin{pmatrix} 2-3\beta & 1-2\beta \\ 1-2\beta & 1-\beta \end{pmatrix}.$$

(3.2.40)

The positivity of Δ implies that $\beta \in [0, 1]$ must satisfy the following two conditions:

$$0 \leq 2 - 3\beta,$$ (3.2.41a)
$$0 \leq -\beta^2 - \beta + 1,$$ (3.2.41b)

from which it follows that $\beta \in [0, (\sqrt{5} - 1)/2]$. On the other hand, we find that

$$\phi_\beta(v) = \begin{pmatrix} 0 & 1 \end{pmatrix} \begin{pmatrix} -\dfrac{1}{\beta} + \dfrac{1}{1-\beta} & -\dfrac{1}{\beta} \\ -\dfrac{1}{\beta} & \dfrac{1}{1-\beta} \end{pmatrix}^{-1} \begin{pmatrix} 0 \\ 1 \end{pmatrix} v^2 = \dfrac{\beta(1-\beta)(2\beta-1)}{\beta^2 + \beta - 1} v^2.$$

(3.2.42)

The graph of $\phi_\beta(1)$ is given in figure 3.2.3. The maximum can again be found by

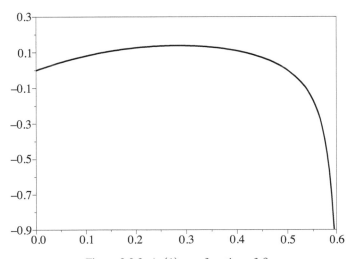

Figure 3.2.3 $\phi_\beta(1)$ as a function of β.

setting the derivative of $\phi_\beta(v)$ with respect to β to zero, which gives us

$$1 - 6\beta + 10\beta^2 - 4\beta^3 - 2\beta^4 = 0. \qquad (3.2.43)$$

Consequently $\beta^* = 0.2826$ and $\phi_{\beta^*}(1) = 0.1383$. Thus $v^* = 2.69$. To see that this value of v is indeed optimal, see figure 3.2.4, which shows the $\{u, y\}$'s in $\mathcal{A}_0$ and $\mathcal{A}_1$. The figure shows $\mathcal{A}_0$, which does not depend on v, and $\mathcal{A}_1$, which rises as v is increased, for $v = 0$, $v^*/2$, and v^*. Clearly $v = v^* = 2.69$ is the smallest value of v for which the two sets do not intersect.

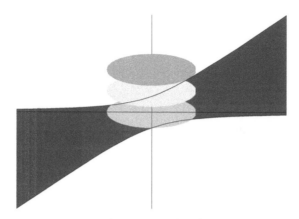

Figure 3.2.4   The sets $\mathcal{A}_0$ and $\mathcal{A}_1$ of $\{u, y\}$ corresponding to the
two models for $v = 0$, $v^*/2$, and $v^* = 2.69$.

Theorem 3.2.1 provides a constructive, two-step method for determining the optimal proper auxiliary signal. It turns out that it is possible to combine the two steps and obtain a one-step method which does not require the explicit computation of V_β and thus is easier to implement.

Lemma 3.2.4 *Suppose (3.2.19) is satisfied. Then λ_β defined in (3.2.20) corresponds to the largest λ for which*

$$\det(\lambda H J_\beta^{-1} H^T - GQ^{-1}G^T) = 0. \qquad (3.2.44)$$

Theorem 3.2.2 *Suppose (3.2.19) is satisfied for all β in a nonempty interval (a, b). Let*

$$\lambda^* = \max_{\beta \in [a,b]} \lambda_\beta, \qquad (3.2.45)$$

where λ_β is the largest λ satisfying (3.2.44). Then a proper auxiliary signal exists if and only if $\lambda^* > 0$. Assume $\lambda^* > 0$. Let β^* be a value of β maximizing in (3.2.45). Let

$$K = G^l H J_{\beta^*}^{-1} H^T, \qquad (3.2.46)$$

where G^l is any left inverse of G. Let ζ be any nonzero vector satisfying

$$(\lambda^* H J_{\beta^*}^{-1} H^T - GQ^{-1}G^T)\zeta = 0 \qquad (3.2.47)$$

and let

$$\alpha = \frac{1}{\sqrt{\lambda^* \zeta^T K^T Q K \zeta}}, \quad \zeta^* = \alpha \zeta. \tag{3.2.48}$$

Then an optimal auxiliary signal is

$$v^* = K \zeta^*. \tag{3.2.49}$$

Example 3.2.4 *Consider the following two systems:*

$$\begin{pmatrix} 0 & 0 \\ 0 & -2 \end{pmatrix} v + y = \nu_0, \tag{3.2.50a}$$

$$\nu_0^T \nu_0 < 1 \tag{3.2.50b}$$

and

$$\begin{pmatrix} -1 & 0 \\ 0 & 0 \\ 0 & 1 \end{pmatrix} v + \begin{pmatrix} 1 & 0 \\ 0 & 1 \\ 0 & 0 \end{pmatrix} y = \nu_1, \tag{3.2.51a}$$

$$\nu_1^T \operatorname{diag}(1, 1, -1) \nu_1 < 1. \tag{3.2.51b}$$

It is straightforward to show that in this case we have

$$H = \begin{pmatrix} 0 & 0 & 0 & 0 & 1 \\ 0 & 1 & 0 & -1 & 0 \\ 1 & 0 & -1 & 0 & 0 \end{pmatrix}, \tag{3.2.52}$$

$$G = \begin{pmatrix} 0 & 1 \\ 0 & -2 \\ 1 & 0 \end{pmatrix}, \tag{3.2.53}$$

$$J_\beta = \operatorname{diag}(\beta, \beta, 1 - \beta, 1 - \beta, \beta - 1). \tag{3.2.54}$$

To determine the a and b of theorem 3.2.2, we use (3.2.19) to see that

$$H_\perp^T J_\beta H_\perp = \frac{1}{2} I > 0. \tag{3.2.55}$$

Thus $(a, b) = (0, 1)$. Figure 3.2.5 shows λ_β as a function of β using the method presented in lemma 3.2.4. We obtain $\lambda^* = 0.5625$ and $\beta^* = 0.625$. Then using the method presented in theorem 3.2.2, we get

$$v^* = \begin{pmatrix} 0 \\ 4 \\ 3 \end{pmatrix}. \tag{3.2.56}$$

The two realizable sets $\mathcal{A}_0(v)$ and $\mathcal{A}_1(v)$ of $\{u, y\}$ corresponding to this value of v are shown in figure 3.2.6.

We end this section with an example that illustrates the importance of the condition (3.2.19).

Example 3.2.5 *Consider the following two systems:*

$$\begin{pmatrix} -1 \\ 0 \end{pmatrix} v + \begin{pmatrix} 0 \\ 1 \end{pmatrix} u + \begin{pmatrix} 1 \\ 0 \end{pmatrix} y = 7\nu_0, \tag{3.2.57a}$$

$$\nu_0^T \nu_0 < 1 \tag{3.2.57b}$$

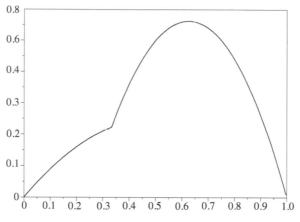

Figure 3.2.5 λ_β as a function of β.

and

$$\begin{pmatrix} 0 \\ 1 \end{pmatrix} u + \begin{pmatrix} 1 \\ 0 \end{pmatrix} y = \nu_1, \tag{3.2.58a}$$

$$\nu_1^T \operatorname{diag}(1, -1)\nu_1 < 1. \tag{3.2.58b}$$

If we disregard condition (3.2.19), λ_β can be computed for all $\beta \in [0, 1]$ as shown in figure 3.2.7. The "wrong" value of β^* can then be obtained by finding the maximum of λ_β, giving $\beta^* = 7/8$. The corresponding v^* is 8. But this v^* does not separate the two sets, as can be seen in figure 3.2.8. Taking into account the condition (3.2.19), it can be shown that the maximum should be taken only over the interval $\beta \in [0.98, 1]$. The maximum is clearly attained at $\beta^* = 0.98$, giving $v^* = 10$, which is an optimal auxiliary signal as illustrated in figure 3.2.9.

3.2.2 On-line Detection Test

Once the auxiliary signal v is constructed, it can be used for on-line failure detection. Unlike the construction of v, which is done off-line and thus can be computationally intensive, the on-line computation must be as efficient as possible.

Standard Solution

We have already seen in the previous chapter that the realizability of a given input-output pair $\{u, y\}$ by a given model can be tested using a single inequality test. In particular, we have the following lemma.

Lemma 3.2.5 *Suppose model i defined in (3.2.2a) satisfies*

$$H_{i\perp}^T J_i H_{i\perp} > 0. \tag{3.2.59}$$

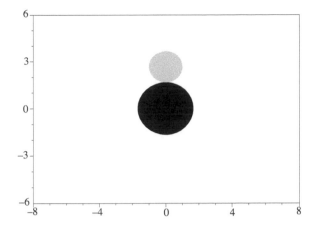

Figure 3.2.6   Realizable sets of $\{u, y\}$, $\mathcal{A}_0(v)$ and $\mathcal{A}_1(v)$, corresponding to $v = v^*$ in example 3.2.4.

Then $\{v, u, y\}$ is realizable by model i, i.e., $\{u, y\} \in \mathcal{A}_i(v)$, if and only if

$$\begin{pmatrix} v \\ u \\ y \end{pmatrix}^T \Omega_i \begin{pmatrix} v \\ u \\ y \end{pmatrix} < 1, \tag{3.2.60}$$

where

$$\Omega_i = \begin{pmatrix} X_i & Y_i & Z_i \end{pmatrix}^T (H_i J_i^{-1} H_i^T)^{-1} \begin{pmatrix} X_i & Y_i & Z_i \end{pmatrix}. \tag{3.2.61}$$

This result is an immediate consequence of theorem 2.6.1.

The detection tests (3.2.60), $i = 0, 1$, can be used once u and y are measured, to decide whether or not a failure has occurred. In fact, since by assumption the true system corresponds necessarily to one of the two models and the two sets $\mathcal{A}_0(v)$ and $\mathcal{A}_1(v)$ are disjoint, it suffices to use one of the two tests because one and only one of the two tests can be satisfied for a given $\{u, y\}$.

Hyperplane Test

It turns out that the sets $\mathcal{A}_0(v)$ and $\mathcal{A}_1(v)$ are convex in many situations of interest. This, of course, is not always the case, as is illustrated by example 3.2.4. The following lemma which is the direct consequence of (3.2.60) characterizes the convex case.

Lemma 3.2.6 *The set $\mathcal{A}_i(v)$ is bounded and convex for all v if and only if*

$$\begin{pmatrix} Y_i & Z_i \end{pmatrix}^T (H_i J_i^{-1} H_i^T)^{-1} \begin{pmatrix} Y_i & Z_i \end{pmatrix} \geq 0. \tag{3.2.62}$$

When the two sets are convex, we can construct a more efficient on-line detection test. Using the fact that two disjoint convex sets can be separated by a hyperplane,

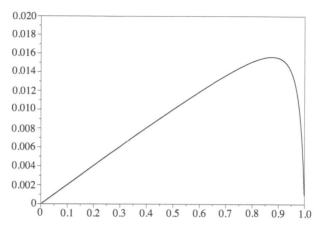

Figure 3.2.7 λ_β disregarding condition (3.2.19).

we can construct a detection test based on geometry for deciding on which side of a hyperplane $\{u, y\}$ is situated. Provided convexity holds, hyperplane tests can be constructed for any proper v. If v is the minimum proper auxiliary signal, then the hyperplane test can be expressed as follows:

$$h^T \left(\begin{pmatrix} u \\ y \end{pmatrix} - \begin{pmatrix} u^* \\ y^* \end{pmatrix} \right) \gtreqless 0, \qquad (3.2.63)$$

where $\{u^*, y^*\}$ is in the intersection of the closures of $\mathcal{A}_0(v)$ and $\mathcal{A}_1(v)$, and h is a vector orthogonal to the tangents to the boundaries of $\mathcal{A}_0(v)$ and $\mathcal{A}_1(v)$ at $\{u^*, y^*\}$. If one of the $\mathcal{A}_i(v)$ is strictly convex, then $\{u^*, y^*\}$ is unique.

The key point here is that h, u^*, and y^* do not depend on u and y. They depend only on v^*. Thus $\{h, u^*, y^*\}$ can be computed off-line.

We know the H_i are full row rank by assumption. In most applications we would have that H_i would be invertible. If an H_i is not full column rank, then there is a projection of ν_i that is zero and can be dropped to give an invertible H_i. However, this requires altering the J_i. To see this, consider the following example.

Example 3.2.6 *Suppose that* $H_i = \begin{pmatrix} 1 & 0 \end{pmatrix}$ *and that* $J_i = \begin{pmatrix} 1 & 1/\sqrt{2} \\ 1/\sqrt{2} & 1 \end{pmatrix}$. *Then*

$$\eta^T J_i \eta = \eta_1^2 + \sqrt{2}\eta_1 \eta_2 + \eta_2^2 < 1. \qquad (3.2.64)$$

If we set $\eta_2 = 0$ *in (3.2.64) we get the constraint* $\eta_1^2 < 1$. *However,* $\eta_1 = 1.1, \eta_2 = -1$ *satisfies (3.2.64). Thus we can set* $\eta_2 = 0$ *but we must then use* $\alpha \eta_1^2 < 1$ *for* $\alpha > 1$ *instead of* $\eta_1^2 < 1$.

Theorem 3.2.3 *Consider model (3.2.2a) for* $i = 0, 1$, *and let* W_i *be defined as in (3.2.13). Suppose a separating hyperplane (3.2.63) exists for a minimum proper* v. *Then one such hyperplane is given by*

$$h = \left(\begin{pmatrix} W_0 & -W_1 \end{pmatrix} \begin{pmatrix} Y_0 & Z_0 \\ Y_1 & Z_1 \end{pmatrix} \right)^T \zeta^*, \qquad (3.2.65)$$

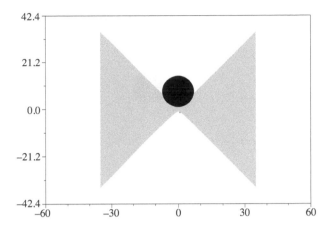

Figure 3.2.8 The wrong v^* obtained by disregarding condition (3.2.19) does not separate the two sets.

where ζ^* is defined in theorem 3.2.2, and

$$\begin{pmatrix} u^* \\ y^* \end{pmatrix} = \begin{pmatrix} Y_0 & Z_0 \\ Y_1 & Z_1 \end{pmatrix}^l \left(\begin{pmatrix} H_0 & 0 \\ 0 & H_1 \end{pmatrix} v^* - \begin{pmatrix} X_0 \\ X_1 \end{pmatrix} v^* \right) \qquad (3.2.66)$$

with

$$v^* = J_\beta^{-1} H^T \zeta^*. \qquad (3.2.67)$$

Proof. We shall prove theorem 3.2.3 when H_i is invertible. We start with the following useful result.

Lemma 3.2.7 *Let z_i be in R^n with $n \geq 2$. Suppose the optimization problem*

$$\min_{z_1, z_2} f(z_1) + g(z_2) \quad \text{subject to} \quad 0 = z_1 - z_2, \qquad (3.2.68)$$

where f and g are differentiable functions, has a finite solution $z_1 = z_2 = e$. Then the sets (level curves) $f(z) = f(e)$ and $g(z) = g(e)$ are tangent to each other at $z = e$. Moreover,

$$\lambda = \frac{\partial f}{\partial z}(e) = -\frac{\partial g}{\partial z}(e) \qquad (3.2.69)$$

is orthogonal to this tangent at $z = e$.

Proof. To prove lemma 3.2.7, note that by assumption $f(z) + g(z)$ has a minimum at e. Thus $f_z(e) + g_z(e) = 0$, which is (3.2.69). But the gradient of a function at a point is always orthogonal to the level surface of the function through that same point. $\qquad \square$

Returning to the proof of theorem 3.2.3, note that λ in (3.2.69) is the Lagrange multiplier associated with the constrained optimization problem (3.2.68). However, without additional assumptions nothing stops λ from being zero.

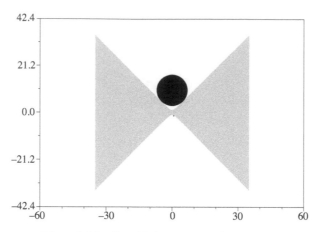

Figure 3.2.9 $v^* = 10$ does separate the two sets.

To see how lemma 3.2.7 can be used here, let us rewrite the optimization problem (3.2.9a) as follows:

$$\sigma(v) = \max_{\beta \in [0,1]} \inf_{\substack{\nu_0, \nu_1, u_0, u_1 \\ y_0, y_1}} (\beta \nu_0^T J_0 \nu_0 + (1 - \beta) \nu_1^T J_1 \nu_1) \qquad (3.2.70a)$$

subject to

$$(X_i \quad Y_i \quad Z_i) \begin{pmatrix} v \\ u_i \\ y_i \end{pmatrix} = H_i \nu_i, \quad i = 0, 1, \qquad (3.2.70b)$$

$$\begin{pmatrix} u_1 \\ y_1 \end{pmatrix} - \begin{pmatrix} u_0 \\ y_0 \end{pmatrix} = 0. \qquad (3.2.70c)$$

Let $\beta^*, u^*, y^*, \nu_0^*,$ and ν_1^* denote a solution to the optimization problem (3.2.9a). It is straightforward to show that $u^*, y^*,$ and

$$\begin{pmatrix} \nu_0^* \\ \nu_1^* \end{pmatrix} = \nu^* \qquad (3.2.71)$$

are given by (3.2.66) and (3.2.67).

The quantity $\sigma(v)$ is defined in (3.2.8) as the infimum over the maximum of two quadratic forms. The global minimum of each of these forms is less than or equal to zero. In fact, it is 0 or $-\infty$. If $\sigma(v) > 0$, then part 3 of lemma 3.9.2 implies that

$$\sigma(v) = \nu_0^{*T} J_0 \nu_0^* = \nu_1^{*T} J_1 \nu_1^*. \qquad (3.2.72)$$

Now consider the optimization problem (3.2.70a), which has the same solutions as (3.2.9a), and rewrite it as follows:

$$\sigma(v) = \inf_{u_0, u_1, y_0, y_1} f(u_0, y_0) + g(u_1, y_1) \qquad (3.2.73a)$$

subject to

$$\begin{pmatrix} u_1 \\ y_1 \end{pmatrix} - \begin{pmatrix} u_0 \\ y_0 \end{pmatrix} = 0, \qquad (3.2.73b)$$

where $f(u_0, y_0)$ and $g(u_1, y_1)$ are given by

$$\beta^* \left(H_0^{-1} \begin{pmatrix} X_0 & Y_0 & Z_0 \end{pmatrix} \begin{pmatrix} v \\ u_0 \\ y_0 \end{pmatrix} \right)^T J_0 \left(H_0^{-1} \begin{pmatrix} X_0 & Y_0 & Z_0 \end{pmatrix} \begin{pmatrix} v \\ u_0 \\ y_0 \end{pmatrix} \right)$$

(3.2.73c)

and

$$(1 - \beta^*) \left(H_1^{-1} \begin{pmatrix} X_1 & Y_1 & Z_1 \end{pmatrix} \begin{pmatrix} v \\ u_1 \\ y_1 \end{pmatrix} \right)^T J_1 \left(H_1^{-1} \begin{pmatrix} X_1 & Y_1 & Z_1 \end{pmatrix} \begin{pmatrix} v \\ u_1 \\ y_1 \end{pmatrix} \right),$$

(3.2.73d)

respectively. Applying lemma 3.2.7 to this problem, we obtain

$$
\begin{aligned}
\lambda &= \beta^* \begin{pmatrix} Y_0^T \\ Z_0^T \end{pmatrix} H_0^{-T} J_0 \nu_0^* - (1 - \beta^*) \begin{pmatrix} Y_1^T \\ Z_1^T \end{pmatrix} H_1^{-T} J_1 \nu_1^* \\
&= \begin{pmatrix} Y_0^T & Y_1^T \\ Z_0^T & Z_1^T \end{pmatrix} \begin{pmatrix} \beta^* H_0^{-T} J_0 & 0 \\ 0 & (\beta^* - 1) H_1^{-T} J_1 \end{pmatrix} \nu^* \\
&= \left(\begin{pmatrix} W_0 & -W_1 \end{pmatrix} \begin{pmatrix} Y_0 & Z_0 \\ Y_1 & Z_1 \end{pmatrix} \right)^T \zeta^*.
\end{aligned}
$$

(3.2.74)

But this is exactly h. Thus h defines the hyperplane tangent to the two sets $\mathcal{A}_0(v)$ and $\mathcal{A}_1(v)$ at a point of contact. Clearly, if a separating hyperplane at this point exists, it must be the one defined by h, and theorem 3.2.3 is proven. □

The advantage of the hyperplane test over the standard test is not significant in the static case. However, the difference becomes significant in the dynamic case as we shall see later.

Example 3.2.7 *To illustrate the computation of the separating hyperplane, consider again the two models in example 3.2.4. In this case we have*

$$W_0 = \begin{pmatrix} 0 & 0 \\ 0 & 1 \\ 1 & 0 \end{pmatrix}, \quad Z_0 = \begin{pmatrix} 1 & 0 \\ 0 & 1 \end{pmatrix}, \quad W_1 = \begin{pmatrix} 0 & 0 & 1 \\ 0 & -1 & 0 \\ -1 & 0 & 0 \end{pmatrix}, \quad Z_1 = \begin{pmatrix} 1 & 0 \\ 0 & 1 \\ 0 & 0 \end{pmatrix}.$$

(3.2.75)

From (3.2.65) and (3.2.66), we obtain

$$h = \begin{pmatrix} 0 \\ -1 \end{pmatrix}, \quad y^* = \begin{pmatrix} 0 \\ \frac{5}{3} \end{pmatrix}.$$

(3.2.76)

The resulting hyperplane is illustrated in figure 3.2.10.

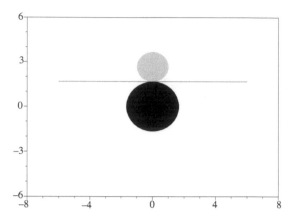

Figure 3.2.10 Hyperplane separating realizable sets $\mathcal{A}_0(v)$ and $\mathcal{A}_1(v)$.

3.3 CONTINUOUS-TIME CASE

We now turn to discussing the important case where the models are differential equations.

As we did in the previous section, we consider a slight generalization of the continuous-time model (2.3.21a)–(2.3.21c) presented in the previous chapter. By using ν to represent ν and z in (2.3.21a)–(2.3.21c), and by breaking up the input u into two inputs u and v, we can include the model (2.3.21a)–(2.3.21c) of chapter 2 in the one we consider here, which is

$$\dot{x}_i = A_i x_i + B_i v + \bar{B}_i u + M_i \nu_i, \tag{3.3.1a}$$

$$E_i y = C_i x_i + D_i v + \bar{D}_i u + N_i \nu_i. \tag{3.3.1b}$$

Here $i = 0, 1$ correspond to the normal and failed system models, respectively. The only conditions on the system matrices are that the N_i's have full row rank and the E_i's have full column rank (for all t in $[0, T]$ if they depend on t).

In this new notation, the constraint (or noise measure) on the initial condition and uncertainties (2.3.24) can be expressed as follows:

$$\mathcal{S}_i(v, s) = x_i(0)^T P_{i0}^{-1} x_i(0) + \int_0^s \nu_i^T J_i \nu_i \, dt < 1, \ \forall \, s \in [0, T], \tag{3.3.2}$$

where the J_i's are signature matrices. As discussed in chapter 2, the bound (3.3.2) allows for both additive and model uncertainty. One major difference in the two cases is that with additive uncertainty we have $J_i = I$ and thus we can just take $s = T$ in (3.3.2).

The assumption is that for failure detection we have access to u and y, given a v, consistent with one of the models. The problem we consider here is to find an optimal v in some sense for which observation of u and y provides enough information to decide from which model u, y have been generated. For example, if the optimization is on the L^2 norm of v, our problem becomes that of finding a v of smallest L^2 norm for which there exist no solutions to (3.3.1a), (3.3.1b), and (3.3.2) for $i = 0$ and 1 simultaneously.

Even though technically the problem in the continuous-time case is quite different from the one we considered in the static case, the overall approach is very similar. We start by defining $\sigma(v, s)$, which is the counterpart of $\sigma(v)$ in the static case.

Note that since the N_i's are full row rank we have that for any L^2 functions v, u, and y, there exist L^2 functions ν_i satisfying (3.3.1a),(3.3.1b). Thus the nonexistence of a solution to (3.3.1a), (3.3.1b), and (3.3.2) is equivalent to

$$\sigma(v, s) \geq 1 \quad \text{for some } s, \tag{3.3.3a}$$

where

$$\sigma(v, s) = \inf_{\substack{\nu_0, \nu_1, u, y \\ x_0, x_1}} \max(\mathcal{S}_0(v, s), \mathcal{S}_1(v, s)) \tag{3.3.3b}$$

subject to (3.3.1a),(3.3.1b) for $i = 0, 1$.

Following the same steps that we did in the static case, we can express $\sigma(v, s)$ as follows:

$$\sigma(v, s) = \max_{\beta \in [0,1]} \phi_\beta(v, s), \tag{3.3.4a}$$

where

$$\phi_\beta(v, s) = \inf_{\substack{\nu_0, \nu_1, u, y \\ x_0, x_1}} \beta \mathcal{S}_0(v, s) + (1 - \beta)\mathcal{S}_1(v, s) \tag{3.3.4b}$$

subject to (3.3.1a),(3.3.1b), $i = 0, 1$.

We start by considering the case where there is no u, i.e., all the inputs are used as an auxiliary signal for failure detection. Then by using the notation

$$x = \begin{pmatrix} x_0 \\ x_1 \end{pmatrix}, \nu = \begin{pmatrix} \nu_0 \\ \nu_1 \end{pmatrix}, A = \begin{pmatrix} A_0 & 0 \\ 0 & A_1 \end{pmatrix}, B = \begin{pmatrix} B_0 \\ B_1 \end{pmatrix}, \tag{3.3.5a}$$

$$M = \begin{pmatrix} M_0 & 0 \\ 0 & M_1 \end{pmatrix}, \tag{3.3.5b}$$

$$D = F_0 D_0 + F_1 D_1, \ C = \begin{pmatrix} F_0 C_0 & F_1 C_1 \end{pmatrix}, \ N = \begin{pmatrix} F_0 N_0 & F_1 N_1 \end{pmatrix}, \tag{3.3.5c}$$

$$P_\beta^{-1} = \begin{pmatrix} \beta P_{0,0}^{-1} & 0 \\ 0 & (1 - \beta) P_{1,0}^{-1} \end{pmatrix}, J_\beta = \begin{pmatrix} \beta J_0 & 0 \\ 0 & (1 - \beta) J_1 \end{pmatrix}, \tag{3.3.5d}$$

where

$$F = \begin{pmatrix} F_0 & F_1 \end{pmatrix} = \begin{pmatrix} E_0 \\ E_1 \end{pmatrix}^\perp, \tag{3.3.5e}$$

we can reformulate the problem (3.3.4) as follows:

$$\phi_\beta(v, s) = \inf_{\nu, x} x(0)^T P_\beta^{-1} x(0) + \int_0^s \nu^T J_\beta \nu \, dt \tag{3.3.6}$$

subject to

$$\dot{x} = Ax + Bv + M\nu, \tag{3.3.7a}$$

$$0 = Cx + Dv + N\nu. \tag{3.3.7b}$$

Note that because the N_i are full row rank, but usually not square, it is usually the case that N of (3.3.5c) has a nontrivial nullspace. That is, $N_\perp \neq 0$. Before proceeding it is helpful to note the following useful fact.

Lemma 3.3.1 *Suppose that S_0, S_1 are two symmetric matrices. Let $S_\beta = \beta S_0 + (1 - \beta)S_1$.*

1. *If $\beta_1 < \beta_2$ and $S_{\beta_i} \geq 0$, then $S_\beta \geq 0$ for $\beta \in [\beta_1, \beta_2]$.*

2. *If $S_{\hat{\beta}} > 0$ for some $\hat{\beta}$, then there are β_1, β_2 such that $\beta_1 < \beta_2$ and $S_\beta > 0$ for $\beta \in (\beta_1, \beta_2)$, $S_{\beta_i} \geq 0$, and S_β is indefinite if $\beta \notin [\beta_1, \beta_2]$. (The end points β_1, β_2 can be $\pm\infty$.)*

Proof. Observe that for $\rho \in [0\ 1]$, we have $\rho S_{\beta_1} + (1-\rho)S_{\beta_2} = S_{\rho\beta_1 + (1-\rho)\beta_2}$. □

A similar statement holds if we consider quadratic forms involving inner products. It turns out that in order to get a proper signal we need that $\phi_\beta(v, s) > -\infty$. The next two lemmas are useful for making a number of observations.

Lemma 3.3.2 *Let $\mathcal{B}$ be the set of all β such that $\phi_\beta(v, s) > -\infty$ for all $s \leq T$. Suppose that N and J_β are constant. If $\beta \in \mathcal{B}$, then $N_\perp^T J_\beta N_\perp \geq 0$.*

Proof. Let

$$Q_{\beta,s}(x_0, \nu) = x(0)^T P_\beta^{-1} x(0) + \int_0^s \nu^T J_\beta \nu \, dt. \qquad (3.3.8)$$

Then $Q_{\beta,s}$ is a quadratic form on the subspace of $R^n \times L^2$ defined by the linear constraints (3.3.7). This subset is of the form $Rv + \mathcal{N}$ where R is a a bounded linear transformation and $\mathcal{N}$ is a subspace of $R^n \times L^2$. If $Q_{\beta,s}(x_0, \nu) < 0$ for some $(x_0, \nu) \in \mathcal{N}$, then by taking larger multiples of (x_0, ν) we would see that $\phi_\beta(v, s) = -\infty$ and $\beta \notin \mathcal{B}$. Thus $Q_{\beta,s}$ is positive semidefinite on $\mathcal{N}$ if $\beta \in \mathcal{B}$. Looking at (3.3.7) with $v = 0$ we get in the constant coefficient case that $\mathcal{N}$ is given by

$$\nu(t) = -N^\dagger C e^{At} x(0) + (I - N^\dagger C \mathcal{L} M) N_\perp \eta(t) \qquad (3.3.9)$$

where $\mathcal{L}f$ is the solution of $\dot{x} = Ax + f$, $x(0) = 0$. (e^{At} is replaced by the fundamental solution matrix if A is time varying.) The function η is arbitrary in $L^2[0, s]$. Suppose that it is not true that $N_\perp^T J_\beta N_\perp \geq 0$. Let ψ be such that $\psi^T N_\perp^T J_\beta N_\perp \psi < 0$. Let $\tilde{\nu} = (I - N^\dagger C \mathcal{L} M) N_\perp \psi$. But viewed as an operator on $L^2[0, s]$ we have $\lim_{s\to 0+} \|\mathcal{L}\| = 0$. Thus for small enough s we get that $Q_{\beta,s}$ is not nonnegative, which is a contradiction. □

Lemma 3.3.3 *Suppose the model uncertainty is in the form (2.3.20). Then there is a $T_o > 0$ such that for all $s \leq T_o$, we have $\phi_\beta(v, s) > 0$ for all $0 < \beta < 1$ and hence $\mathcal{B} = [0, 1]$.*

Proof. The proof consists of observing that $\nu^T J_i \nu = \|\nu\|^2 - \|z\|^2$ and that on the set where $v = 0$ we have that z depends on an integral operator $\mathcal{L}_i$ whose norm goes to zero as $T \to 0^+$. Thus we can get $\|z\| \leq k\|\nu\|^2$ where $k < 1$ for small enough T. □

A interesting side observation is that the output sets are strictly convex if $T < T_o$. This follows from the set of ν satisfying (3.3.2) being a bounded and hence convex set in L^2.

Lemma 3.3.4 $\mathcal{B}$ *is a subinterval of* $[0, 1]$. *If in addition we have that* $Q_{\beta,s}$ *from (3.3.8) is positive semidefinite on* $\mathcal{N}$ *for all* s, *then* $\mathcal{B}$ *is independent of* v.

Proof. $\mathcal{N}$ is a subspace of $R^n \times L^2$ and is independent of β. Let $\mathcal{B}(v, s) = \{\beta : \phi_\beta(v, s) > -\infty\}$. Suppose that $\phi_\beta(v, s) > -\infty$ and $\phi_{\hat{\beta}}(v, s) > -\infty$. Observe that $Q_{\rho\beta+(1-\rho)\hat{\beta},s} = \rho Q_{\beta,s} + (1 - \rho)Q_{\hat{\beta},s}$. Thus $\phi_{\rho\beta+(1-\rho)\hat{\beta}}(v, s) \geq (1 - \rho)\phi_\beta(v, s) + \rho\phi_{\hat{\beta}}(v, s) > -\infty$. Hence $\mathcal{B}(v, s)$ is an interval. Suppose that $Q_{\beta,s}$ is positive semidefinite on $\mathcal{N}$. Then $\phi_\beta(v, s) > -\infty$. This shows that $\mathcal{B}$ is independent of test signal v. $\qquad\square$

The preceding discussion shows that the three assumptions that $N_\perp$ is nonempty, $N_\perp^T J_\beta N_\perp > 0$ for some β, and $\mathcal{B}$ is not empty are not very restrictive and usually hold in practice. Accordingly we will assume these properties throughout the remainder of our development.

Theorem 3.3.1 *Suppose for some* $\beta \in [0, 1]$ *that*

1. The matrices N *and* J_β *satisfy*

$$N_\perp^T J_\beta N_\perp > 0, \quad \forall\, t \in [0, T]; \tag{3.3.10}$$

2. The Riccati equation

$$\dot{P} = (A - S_\beta R_\beta^{-1} C)P + P(A - S_\beta R_\beta^{-1} C)^T - PC^T R_\beta^{-1} CP$$
$$+ Q_\beta - S_\beta R_\rho^{-1} S_\beta^T, \qquad P(0) = P_\beta, \tag{3.3.11}$$

where

$$\begin{pmatrix} Q_\beta & S_\beta \\ S_\beta^T & R_\beta \end{pmatrix} - \begin{pmatrix} M \\ N \end{pmatrix} J_\beta^{-1} \begin{pmatrix} M \\ N \end{pmatrix}^T, \tag{3.3.12}$$

has a solution on $[0, T]$.

Then $\beta \in \mathcal{B}$.

This result follows easily from theorem 2.6.2.

It turns out, and it is easy to verify, that condition (3.3.10) holds for all the systems we have considered in the previous chapter, and in particular in section 2.3. We shall assume from here on that there exists at least one β for which the two conditions of theorem 3.3.1 are satisfied.

3.3.1 Construction of an Optimal Proper Auxiliary Signal

For the sake of simplicity of presentation, we start by assuming that the optimality criterion used for the auxiliary signal is just that of minimizing its L^2 norm. More general norms are considered later in section 3.5. Thus the problem to be solved here is

$$\min_v \|v\|_T^2 \quad \text{subject to} \quad \max_{\substack{\beta \in [0,1] \\ s \in [0,T]}} \phi_\beta(v, s) \geq 1, \tag{3.3.13}$$

where

$$\|v\|_\tau^2 = \int_0^\tau |v|^2 \, dt. \tag{3.3.14}$$

The reason we take the max over s is that the maximum value of $\phi_\beta(v, s)$ does not always occur at $s = T$. (It does in most cases though.)

Suppose the maximum value occurs at $\hat{s}$ where $\hat{s} < T$. Then thanks to the causal dependence of ϕ_β on v, the solution v of (3.3.13) is identically zero from $\hat{s}$ to T, so that $\|v\|_{\hat{s}} = \|v\|_T$. Then without changing the problem, we can replace $\|v\|_T$ by $\|v\|_{\hat{s}}$ in (3.3.13) and consider v only on $[0, \hat{s}]$.

As we did in the static case, we reformulate the optimization problem as follows:

$$\lambda_{\beta,s} = \max_{v \neq 0} \frac{\phi_\beta(v, s)}{\|v\|_s^2}. \tag{3.3.15}$$

This problem is reformulated as

$$\max_v \phi_\beta(v, s) - \lambda \|v\|_s^2 = \max_v \inf_{\nu,x} x(0)^T P_\beta^{-1} x(0) + \int_0^s \nu^T J_\beta \nu - \lambda |v|^2 \, dt \tag{3.3.16}$$

subject to (3.3.7). The value of $\lambda_{\beta,s}$ corresponds then to the infimum of all the λ's for which the solution of this optimization problem is unique and the corresponding cost finite (zero).

Theorem 3.3.2 *Suppose the two conditions of theorem 3.3.1 are satisfied. Then $\lambda_{\beta,s}$ is the infimum of the set of all λ for which, on the interval $[0, s]$,*

  1. The matrices D, R_β satisfy

$$\lambda I - D^T R_\beta^{-1} D > 0; \tag{3.3.17}$$

  2. The Riccati equation

$$\dot{P} = (A - S_{\lambda,\beta} R_{\lambda,\beta}^{-1} C) P + P(A - S_{\lambda,\beta} R_{\lambda,\beta}^{-1} C)^T - PC^T R_{\lambda,\beta}^{-1} C P$$
$$+ Q_{\lambda,\beta} - S_{\lambda,\beta} R_{\lambda,\beta}^{-1} S_{\lambda,\beta}^T, \quad P(0) = P_\beta, \tag{3.3.18}$$

  has a solution, where

$$\begin{pmatrix} Q_{\lambda,\beta} & S_{\lambda,\beta} \\ S_{\lambda,\beta}^T & R_{\lambda,\beta} \end{pmatrix} = \begin{pmatrix} M & B \\ N & D \end{pmatrix} \begin{pmatrix} J_\beta & 0 \\ 0 & -\lambda I \end{pmatrix}^{-1} \begin{pmatrix} M & B \\ N & D \end{pmatrix}^T. \tag{3.3.19}$$

Proof. Thanks to theorem 2.6.2 and its corollary, we have that

$$\phi_\beta(v, s) = \int_0^s \mu^T R_\beta^{-1} \mu \, dt, \tag{3.3.20}$$

where μ is given by

$$\dot{\hat{x}} = \bar{A}\hat{x} + \bar{B}v, \quad \hat{x}(0) = 0, \tag{3.3.21a}$$
$$\mu = C\hat{x} + Dv, \tag{3.3.21b}$$

where

$$\bar{A} = A - S_\beta R_\beta^{-1} C - PC^T R_\beta^{-1} C, \tag{3.3.22}$$
$$\bar{B} = B - (S_\beta + PC^T) R_\beta^{-1} D. \tag{3.3.23}$$

Thus the optimization problem (3.3.16) can be expressed as follows:

$$\max_v \int_0^s \mu^T R_\beta^{-1} \mu - \lambda |v|^2 \, dt. \tag{3.3.24}$$

This optimization problem can be reformulated as follows in order to use theorem 2.6.2

$$\min_{\mu,v} \int_0^s \begin{pmatrix} \mu \\ v \end{pmatrix}^T \begin{pmatrix} -R_\beta^{-1} & 0 \\ 0 & \lambda \end{pmatrix} \begin{pmatrix} \mu \\ v \end{pmatrix} dt \tag{3.3.25}$$

subject to

$$\dot{\hat{x}} = \bar{A}\hat{x} + \begin{pmatrix} 0 & \bar{B} \end{pmatrix} \begin{pmatrix} \mu \\ v \end{pmatrix}, \tag{3.3.26a}$$

$$0 = C\hat{x} + \begin{pmatrix} -I & D \end{pmatrix} \begin{pmatrix} \mu \\ v \end{pmatrix}, \tag{3.3.26b}$$

with $\hat{x}(0) = 0$.

The first condition given by theorem 2.6.2 is

$$\begin{pmatrix} -I & D \end{pmatrix}_\perp^T \begin{pmatrix} -R_\beta^{-1} & 0 \\ 0 & \lambda \end{pmatrix} \begin{pmatrix} -I & D \end{pmatrix}_\perp > 0, \tag{3.3.27}$$

which since

$$\begin{pmatrix} D^T & I \end{pmatrix} = \begin{pmatrix} -I & D \end{pmatrix}_\perp^T \tag{3.3.28}$$

is equivalent to (3.3.17).

The second condition is the existence of a solution to the Riccati equation

$$\dot{\Pi} - (\bar{A} - SR^{-1}C)\Pi + \Pi(\bar{A} - SR^{-1}C)^T - \Pi C^T R^{-1} C \Pi$$
$$+ Q - SR^{-1}S^T, \qquad \Pi(0) = 0, \tag{3.3.29}$$

where

$$\begin{pmatrix} Q & S \\ S^T & R \end{pmatrix} = \begin{pmatrix} 0 & \bar{B} \\ -I & D \end{pmatrix} \begin{pmatrix} -R_\beta^{-1} & 0 \\ 0 & \lambda I \end{pmatrix}^{-1} \begin{pmatrix} 0 & \bar{B} \\ -I & D \end{pmatrix}^T. \tag{3.3.30}$$

The Riccati equation (3.3.18) is then obtained by subtracting (3.3.29) from (3.3.11) and then letting $P - \Pi$ be the P of (3.3.18). $\qquad \square$

We do not pay much attention to the condition (3.3.17) in what follows, because this condition is satisfied when the problem comes from the uncertain models considered in chapter 2 except in a very special case. The D matrix here contains the D and H matrices in (2.3.21) for the two models. The presence of the H matrices is never a problem because the corresponding part in the matrix $N J_\beta^{-1} N^T$ is negative definite and thus cannot lead to the violation of (3.3.17). As for the D's, it is their difference that counts. Thus the only case where the condition (3.3.17) may play a role is when the two models (2.3.21) have different D matrices. This is a very special case because usually uncertain models do not have direct feedthrough from v to y, and even if they do, it is usually identical. Thus the condition (3.3.17) has to be considered only when the failure occurs in the feedthrough terms.

Theorem 3.3.2 shows that $\lambda_{\beta,s}$ is an increasing function of s and thus it allows us to compute

$$\lambda_\beta = \max_{s \leq T} \lambda_{\beta,s} \tag{3.3.31}$$

by simply performing a "λ iteration" by testing for the existence of a solution P to (3.3.18) over $[0, T]$. Then λ^* and β^* are obtained from

$$\lambda^* = \max_{\beta \subset \mathcal{B}}, \lambda_\beta \tag{3.3.32}$$

which gives the separability index

$$\gamma^* = \sqrt{\lambda^*}. \tag{3.3.33}$$

The values of λ^* and β^* are later used to compute the optimal auxiliary signal v^*.

If we are considering only additive noise, then λ_β is defined for all β in $(0, 1)$. However, as noted, this may not be the case with model uncertainty. Figure 3.3.1 shows the graph of λ_β for a randomly generated problem with model uncertainty.

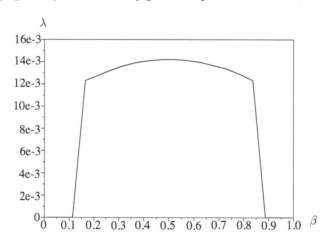

Figure 3.3.1 Graph of λ_β for a typical problem.

Note that the Riccati equation (3.3.18) for $\lambda = \lambda^*$ and $\beta = \beta^*$ does not necessarily have an escape time T; its solution may diverge before the end of the interval. We denote the actual escape time T^*; in most cases it is equal to T. However, in general, we have only that $T^* \leq T$.

Lemma 3.3.5 *The two-point boundary value system*

$$\frac{d}{dt}\begin{pmatrix} x \\ \zeta \end{pmatrix} = \begin{pmatrix} \Omega_{11} & \Omega_{12} \\ \Omega_{21} & \Omega_{22} \end{pmatrix} \begin{pmatrix} x \\ \zeta \end{pmatrix} \tag{3.3.34}$$

with boundary conditions

$$x(0) = P_{\beta^*}\zeta(0), \tag{3.3.35a}$$

$$\zeta(\tau) = 0, \tag{3.3.35b}$$

where

$$\Omega_{11} = -\Omega_{22}^T = A - S_{\lambda^*,\beta^*} R_{\lambda^*,\beta^*}^{-1} C, \qquad (3.3.36a)$$

$$\Omega_{12} = Q_{\lambda^*,\beta^*} - S_{\lambda^*,\beta^*} R_{\lambda^*,\beta^*}^{-1} S_{\lambda^*,\beta^*}^T, \qquad (3.3.36b)$$

$$\Omega_{21} = C^T R_{\lambda^*,\beta^*}^{-1} C \qquad (3.3.36c)$$

is well posed for $\tau < T^*$ but it is not well posed, i.e., it has nontrivial solutions (x^*, ζ^*), for $\tau = T^*$.

See section 3.9.3 for a proof of lemma 3.3.5.

Theorem 3.3.3 *An optimal auxiliary signal is*

$$v^* = \alpha((-B + S_{\lambda^*,\beta^*} R_{\lambda^*,\beta^*}^{-1} D)^T \zeta + D^T R_{\lambda^*,\beta^*}^{-1} Cx) \qquad (3.3.37)$$

where α is a constant such that $\|v^*\| = 1/\gamma^*$.

Proof. An optimal auxiliary signal is obtained by setting to zero the first variation associated with the optimization problem (3.3.16). Denoting the Lagrange multipliers associated with the constraints (3.3.7) by ζ and μ, respectively, we obtain

$$\dot{x} = Ax + Bv + M\nu, \qquad (3.3.38a)$$

$$0 = Cx + Dv + N\nu, \qquad (3.3.38b)$$

$$\dot{\zeta} = -A^T \zeta + C^T \mu, \qquad (3.3.38c)$$

$$J_\beta \nu = M^T \zeta - N^T \mu, \qquad (3.3.38d)$$

$$-\lambda v = B^T \zeta - D^T \mu. \qquad (3.3.38e)$$

After replacing λ and β respectively with λ^* and β^*, this implies

$$\begin{pmatrix} N & D & 0 \\ J_{\beta^*} & 0 & N^T \\ 0 & \lambda^* I & -D^T \end{pmatrix} \begin{pmatrix} \nu \\ v \\ \mu \end{pmatrix} = \begin{pmatrix} Cx \\ M^T \zeta \\ B^T \zeta \end{pmatrix}, \qquad (3.3.39)$$

from which we get

$$v = ((-B + S_{\lambda^*,\beta^*} R_{\lambda^*,\beta^*}^{-1} D)^T \zeta + D^T R_{\lambda^*,\beta^*}^{-1} Cx)/\lambda, \qquad (3.3.40a)$$

$$\mu = -R_{\lambda^*,\beta^*}^{-1} (\lambda^* Cx + S_{\lambda^*,\beta^*}^T \zeta), \qquad (3.3.40b)$$

$$\nu = J_{\beta^*}^{-1} (M^T \zeta - N^T R_{\lambda^*,\beta^*}^{-1} (S_{\lambda^*,\beta^*}^T \zeta + Cx)). \qquad (3.3.40c)$$

The system (3.3.34) is then obtained by substituting these expressions for v, μ, and ν in (3.3.38a)–(3.3.38e). The formula (3.3.37) is just a renormalization of equation (3.3.40a). $\qquad \square$

Theorem 3.3.3 is an extension of theorem 3.2.2 to the infinite-dimensional case.

Note that if $T^* < T$ we can reduce our test period by setting $T = T^*$ because the additional time is not of any use. It does not allow us to improve the separability index. So from here on we assume that $T = T^*$.

We conclude this section with an example illustrating the effect of adding model uncertainty. The example also has a multi-input auxiliary signal v. The solution is found using the Scilab programs of chapter 6.

Example 3.3.1 *We begin by assuming that model 0 is of the form*

$$\dot{x}_0 = \begin{pmatrix} 0 & -1 \\ 1 & 0 \end{pmatrix} x_0 + \begin{pmatrix} 1 & 0 \\ 0 & 1 \end{pmatrix} v + 10^{-4} \begin{pmatrix} 1 & 0 & 0 \\ 0 & 1 & 0 \end{pmatrix} \nu_0, \qquad (3.3.41a)$$

$$y = \begin{pmatrix} 1 & 2 \end{pmatrix} x_0 + \begin{pmatrix} 0 & 0 & 1 \end{pmatrix} \nu_0. \qquad (3.3.41b)$$

Model 1 is the same except that

$$A_1 = \begin{pmatrix} 0 & 3 \\ -3 & 0 \end{pmatrix}. \qquad (3.3.42)$$

These two models have additive uncertainty. The detection horizon is taken as $[0, 10]$. *The test signal is sought of minimum* L^2 *norm. The uncertainty in both models is bounded by*

$$|x_i(0)|^2 + \int_0^{10} |\nu_i(t)|^2 \, dt < 1. \qquad (3.3.43)$$

This is case 1. In case 2 we also assume there is uncertainty in the A_i *of the form*

$$A_0(\delta) = \begin{pmatrix} 0.4\delta_1 & 1 \\ -1 + 0.4\delta_2 & 0 \end{pmatrix}, \quad A_1(\delta) = \begin{pmatrix} 0.4\delta_3 & 3 \\ -3 + 0.4\delta_4 & 0 \end{pmatrix}. \qquad (3.3.44)$$

Such a perturbation can arise, for example, if the models were for spring mass systems and there was uncertainty in the resistance and spring constant. Using the formalism of section 2.3.3, model 0 in case 2 can be written as

$$\dot{x}_0 = \begin{pmatrix} 0 & -1 \\ 1 & 0 \end{pmatrix} x_0 + \begin{pmatrix} 1 & 0 \\ 0 & 1 \end{pmatrix} v + 10^{-4} \begin{pmatrix} 1 & 0 & 0 \\ 0 & 1 & 0 \end{pmatrix} \nu_0, \qquad (3.3.45a)$$

$$z_0 = \begin{pmatrix} 0.4 & 0 \end{pmatrix} x_0, \qquad (3.3.45b)$$

$$y = \begin{pmatrix} 1 & 2 \end{pmatrix} x_0 + \begin{pmatrix} 0 & 0 & 1 \end{pmatrix} \nu_0. \qquad (3.3.45c)$$

Model 1 is the same except for using A_1 *from (3.3.42) instead of* A_0. *The uncertainty bound for both models is*

$$|x_i(0)|^2 + \int_0^s |\nu_i(t)|^2 - |z_i(t)|^2 \, dt < 1, \quad 0 \le s \le 10. \qquad (3.3.46)$$

Figure 3.3.2 graphs λ_β as a function of β for both cases. The higher graph is for additive uncertainty (case 1). The lower graph is for additive and model uncertainty (case 2). Note that in case 2 the maximum is unique but there are two local maximums.

Figure 3.3.3 plots the entries of v for both cases. The two entries of the optimal v for additive uncertainty are given by the lower-amplitude solid lines. The two entries of the optimal v for additive and model uncertainty in case 2 are given by the higher-amplitude dashed lines. As expected, the test signal in case 2 has to be larger in order to overcome the extra uncertainty in the models. In the model uncertainty case v has to be proper no matter what the model perturbation is. It is interesting to note that in the model uncertainty case v has more of its energy toward the start of the test period. This becomes reasonable when one realizes that the model uncertainty results in larger allowed disturbances as x increases but that $x_i(0)$ is bounded. Thus the size of the z term, and hence the uncertainty, increases with t. It is advantageous for the test signal to try to act earlier.

As is to be expected the shape of v is affected by the way the model uncertainty enters the problem. Suppose that in (3.3.44) we restrict the structure of the noise perturbation to $\delta_2 = 0, \delta_4 = 0$. The λ_β graphs are essentially the same but we get the auxiliary signals in figure 3.3.4. Now that there are some additional restrictions on the model uncertainty, we see a v of reduced norm.

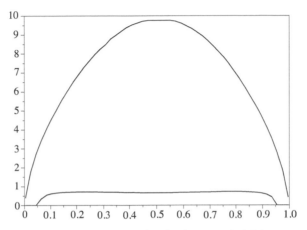

Figure 3.3.2 The function λ_β for example 3.3.1.

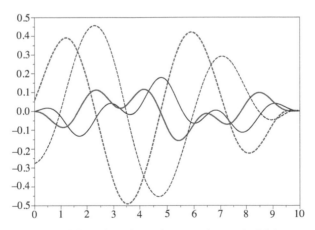

Figure 3.3.3 The v for both cases of example 3.3.1.

3.3.2 On-line Detection Test

Once the auxiliary signal v is constructed, it can be used for on-line failure detection. Unlike the construction of v, which is done off-line and thus can be computationally intensive, the on-line computation burden must be such that real-time implementation can be envisaged for the available hardware.

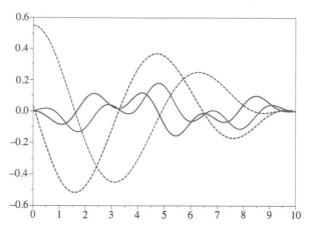

Figure 3.3.4 The v for both cases of example 3.3.1 with restrictions on the δ_i.

Standard Solution

The standard solution to the on-line detection problem here would be simply to use the realizability test derived in the previous chapter for each of the two models. In fact, since by construction

$$\mathcal{A}_0(v^*) \cap \mathcal{A}_1(v^*) = \emptyset \qquad (3.3.47)$$

in a perfect world, we would have to implement only a single realizability test for one of the two models. But in the real world the models are not perfect, and the on-line detection test should be based on both realizability tests.

Another advantage of using two realizability tests is that it could result in faster detection. The reason is that a realizability test can detect "nonrealizability" before the end of the test period T.

The realizability tests for models (3.3.1a),(3.3.1b) are obtained directly from the application of theorem 2.6.2 and its corollary.

Theorem 3.3.4 *Consider model (3.3.1a),(3.3.1b) with the uncertainty model (3.3.2). Suppose that*

1. *Either N_i is invertible for all t or*

$$N_{i\perp}^T J_i N_{i\perp} > 0, \quad \forall\, t \in [0, T]; \qquad (3.3.48)$$

2. *The Riccati equation*

$$\dot{P}_i = (A_i - S_i R_i^{-1} C_i) P_i + P_i (A_i - S_i R_i^{-1} C_i)^T$$
$$- P_i C_i^T R_i^{-1} C_i P_i + Q_i - S_i R_i^{-1} S_i^T, \quad P_i(0) = P_{i0}, \quad (3.3.49)$$

where

$$\begin{pmatrix} Q_i & S_i \\ S_i^T & R_i \end{pmatrix} = \begin{pmatrix} M_i \\ N_i \end{pmatrix} J_i^{-1} \begin{pmatrix} M_i \\ N_i \end{pmatrix}^T \qquad (3.3.50)$$

has a solution on $[0, T]$.

Then a realizability test for model i is

$$\gamma_{i,s}(u,y) < 1, \quad \text{for all } s \in [0,T], \tag{3.3.51}$$

where

$$\gamma_{i,s} = \int_0^s \mu_i^T R_i^{-1} \mu_i \, dt \tag{3.3.52}$$

and μ_i is the output of the following system:

$$\dot{\hat{x}}_i = A_i \hat{x}_i - (S_i + P_i C_i^T) R_i^{-1} \mu_i + B_i v^* + \bar{B}_i u, \tag{3.3.53a}$$

$$\mu_i = C_i \hat{x}_i - (E_i y - D_i v^* - \bar{D}_i u) \tag{3.3.53b}$$

with $\hat{x}_i(0) = 0$.

Hyperplane Test

As in the static case, in some cases it is possible to construct a "hyperplane" test to be used for on-line detection. A separating hyperplane does not always exist, but it does when the output sets $\mathcal{A}_i(v)$ are convex. In the absence of on-line measured inputs u, and under reasonable assumptions, the $\mathcal{A}_i(v)$ are convex. See lemma 2.3.2.

In the continuous-time case, the separating hyperplane test can be expressed as follows:

$$\int_0^T h(t)^T (y(t) - y^*(t)) dt \lesseqgtr 0. \tag{3.3.54}$$

We have noted explicitly the dependence of h and y^* on t to emphasize that they could (and almost always do) depend on time.

Theorem 3.3.5 *Suppose that a separating hyperplane exists and that the two conditions of theorem 3.3.1 are satisfied. Then a separating hyperplane can be characterized as follows:*

$$h = \left(F \begin{pmatrix} E_0 \\ -E_1 \end{pmatrix} \right)^T R_{\lambda^*,\beta^*}^{-1} (Cx + S_{\lambda^*,\beta^*}^T \zeta) \tag{3.3.55}$$

and

$$y^* = \alpha \begin{pmatrix} E_0 \\ E_1 \end{pmatrix}^l \left(\begin{pmatrix} C_0 & 0 \\ 0 & C_1 \end{pmatrix} x - \Psi_{\lambda^*,\beta^*} \left(F^T R_{\lambda^*,\beta^*}^{-1} (Cx + S_{\lambda^*,\beta^*}^T \zeta) \right) \right); \tag{3.3.56}$$

where α is defined in theorem 3.3.3 and

$$\Psi_{\lambda^*,\beta^*} = \begin{pmatrix} J_{\beta^*} & 0 \\ 0 & -\lambda^* I \end{pmatrix}^{-1} \begin{pmatrix} M & B \\ \mathcal{N} & \mathcal{D} \end{pmatrix}^T, \tag{3.3.57}$$

where

$$\mathcal{N} = \begin{pmatrix} N_0 & 0 \\ 0 & N_1 \end{pmatrix}, \quad \mathcal{D} = \begin{pmatrix} D_0 & 0 \\ 0 & D_1 \end{pmatrix}. \tag{3.3.58}$$

Proof. As in the static case, we consider the optimization problem of

$$\inf_{\nu,x,y_0,y_1} x(0)^T P_\beta^{-1} x(0) + \int_0^T \nu^T J_\beta \nu \, dt \tag{3.3.59a}$$

subject to

$$\dot{x} = Ax + Bv + M\nu, \tag{3.3.59b}$$

$$\begin{pmatrix} E_0 & 0 \\ 0 & E_1 \end{pmatrix} \begin{pmatrix} y_0 \\ y_1 \end{pmatrix} = \begin{pmatrix} C_0 & 0 \\ 0 & C_1 \end{pmatrix} x + \begin{pmatrix} D_0 & 0 \\ 0 & D_1 \end{pmatrix} v + \begin{pmatrix} N_0 & 0 \\ 0 & N_1 \end{pmatrix} \nu, \tag{3.3.59c}$$

$$y_0 - y_1 = 0. \tag{3.3.59d}$$

The Lagrange multiplier associated with the constraint (3.3.59d) then gives h. This problem can be solved for any given v but we are, of course, often interested in the case $v = v^*$, although sometimes larger proper signals are used.

Computing the optimality conditions by performing a first-order variation yields the following additional equations:

$$\dot{\zeta} + A^T \zeta + \begin{pmatrix} C_0 & 0 \\ 0 & C_1 \end{pmatrix}^T \omega = 0, \tag{3.3.60a}$$

$$\begin{pmatrix} E_0 & 0 \\ 0 & E_1 \end{pmatrix}^T \omega - \begin{pmatrix} I \\ -I \end{pmatrix} h = 0, \tag{3.3.60b}$$

$$J_\beta \nu - M^T \zeta - \mathcal{N}^T \omega = 0, \tag{3.3.60c}$$

where ζ and ω are the Lagrange multipliers associated with the constraints (3.3.59b) and (3.3.59c), respectively.

From (3.3.60b) it follows that

$$\begin{pmatrix} E_0 \\ E_1 \end{pmatrix}^T \omega = 0. \tag{3.3.61}$$

But thanks to (3.3.5e) we know that the columns of F^T form a basis for the nullspace of $\begin{pmatrix} E_0 \\ E_1 \end{pmatrix}^T$. Thus there exists a σ such that

$$\omega = F^T \sigma. \tag{3.3.62}$$

It is now straightforward to show that

$$\sigma = -R_{\beta^*}^{-1}(Cx + S_{\lambda^*,\beta^*}^T \zeta). \tag{3.3.63}$$

But

$$h = -\begin{pmatrix} E_0^T & -E_1^T \end{pmatrix} F^T \sigma, \tag{3.3.64}$$

which gives (3.3.55). Then (3.3.56) follows easily from (3.3.59c) by letting $y^* = y_0 = y_1$. $\square$

3.3.3 Numerical Issues

All the off-line computations required for the construction of the optimal auxiliary signal and the hyperplane test can be done efficiently using Scilab or Matlab. The computation of $\lambda_{\beta,s}$ requires a "λ iteration" which needs the solution of the Riccati equation (3.3.18). This can be implemented using a standard ordinary differential equation (ODE) solver. Note that for a given β this λ iteration gives the $\max_s \lambda_{\beta,s}$.

The optimization over the scalar β can be done simply by discretizing the set $\mathcal{B}$ (in the worst case $\mathcal{B} = [0,1]$). That gives us β^*.

Finally, the computation of the optimal auxiliary signal and the hyperplane test (if applicable) requires the solution of the two-point boundary value system (3.3.34). This problem is not standard because this system is not well posed. To find the solution, we first need the following result.

Lemma 3.3.6 *For any solution* (x, ζ) *of the boundary value system (3.3.34), we have, over* $[0, T)$,

$$x = P\zeta, \tag{3.3.65}$$

where P is the solution of the Riccati equation (3.3.18) with $\beta = \beta^*$ *and* $\lambda = \lambda^*$.

Proof. Suppose that (x, ζ) satisfies (3.3.34),(3.3.35a). Let $\hat{\zeta}$ be the solution of

$$\dot{\hat{\zeta}} = (\Omega_{21}P + \Omega_{22})\hat{\zeta}, \quad \hat{\zeta}(0) = \zeta(0). \tag{3.3.66}$$

Let $\hat{x} = P\hat{\zeta}$. It is straightforward to verify that $(\hat{x}, \hat{\zeta})$ is a solution of (3.3.34), (3.3.35a). But (x, ζ) and $(\hat{x}, \hat{\zeta})$ are solutions of (3.3.34) which satisfy the same initial condition, and hence $x = \hat{x}$.   □

We have seen that as t goes to T, $P(t)$ diverges so that $\bar{P}(T) = \lim_{t \to T} P(t)^{-1}$ is singular. $\bar{P}(T)$ can be computed by inverting $P(t)$ for a t close to T. However, this may not be numerically reliable. Alternatively, when the computation gets close to T (say, at $t = T - \delta$ for some small δ) and $P(t)$ starts diverging, it switches from the Riccati equation (3.3.18) to the following equivalent Riccati equation (here $\bar{P} = P^{-1}$):

$$-\dot{\bar{P}} = \bar{P}(A - S_{\lambda^*,\beta^*} R_{\lambda^*,\beta^*}^{-1} C) + (A - S_{\lambda^*,\beta^*} R_{\lambda^*,\beta^*}^{-1} C)^T \bar{P} - C^T R_{\lambda^*,\beta^*}^{-1} C$$
$$+ \bar{P}(Q_{\lambda^*,\beta^*} - S_{\lambda^*,\beta^*} R_{\lambda^*,\beta^*}^{-1} S_{\lambda^*,\beta^*}^T)\bar{P}, \quad \bar{P}(T-\delta) = P^{-1}(T-\delta). \tag{3.3.67}$$

This equation can be integrated up to T, yielding a singular $\bar{P}(T)$. We do not use this Riccati equation from the beginning ($t = 0$) since, as noted earlier, $P(t)$ becomes singular somewhere in the middle of the interval, which means that $\bar{P}$ goes through infinity at that point. Thus δ must be chosen small enough to avoid this singularity, but not too small in order to avoid numerical problems with the inversion of $P(T-\delta)$.

Once we have found $\bar{P}(T)$, noting that $\bar{P}(T)x(T) = \zeta(T) = 0$, we can let $x(T)$ be any nonzero vector x_T in the nullspace of $\bar{P}(T)$. We can then find a nonzero solution to (3.3.34) by taking as the boundary (final) condition

$$x(T) = x_T, \quad \zeta(T) = 0. \tag{3.3.68}$$

This system is well posed and has a unique solution. However, since this system is not backward stable, its numerical solution can result in large errors. The way to avoid this problem is to use (3.3.34) with boundary (final) condition (3.3.68) simply to solve for (x, ζ) over a short interval, for example, $[T - \delta, T]$. Then from $T - \delta$ on, use

$$\dot{\zeta} = (-A^T + C^T R^{-1}_{\lambda^*,\beta^*} CP + C^T R^{-1}_{\lambda^*,\beta^*} S^T_{\lambda^*,\beta^*}) \zeta \qquad (3.3.69)$$

to solve for ζ down to zero. This relation is obtained easily from (3.3.65). The optimal auxiliary signal is then given by (3.3.37).

A Scilab implementation of this procedure is given in section 6.2 of chapter 6.

3.4 CASE OF ON-LINE MEASURED INPUT

Up until now in this chapter we have made the assumption that the system has no on-line measured inputs u. If it does, the problem becomes more complicated. The u input is in many respects similar to the output y: its value is not known in advance but it is available on-line. However, u cannot be eliminated the way y was eliminated because u appears in the dynamics part of the equations. In this section we shall address this problem with the additional assumption that the system matrices are constant. We can relax this assumption but then the formulas and notation become much more complicated. In particular, coordinate changes such as (3.4.12) will then be time varying. This in turn causes derivatives of these coordinate changes to appear in (3.4.13) and the subsequent equations.

Proceeding exactly as we did in the previous section, from (3.3.1a) and (3.3.1b), keeping u in the equations, we obtain a slight generalization of the optimization problem (3.3.16):

$$\max_v \inf_{\nu,x,u} x(0)^T P_\beta^{-1} x(0) + \int_0^T \nu^T J_\beta \nu - \lambda |v|^2 \, dt \qquad (3.4.1a)$$

subject to

$$\dot{x} = Ax + Bv + M\nu + \bar{B}u, \qquad (3.4.1b)$$

$$0 = Cx + Dv + N\nu + \bar{D}u, \qquad (3.4.1c)$$

where

$$\bar{B} = \begin{pmatrix} \bar{B}_0 \\ \bar{B}_1 \end{pmatrix}, \quad \bar{D} = F_0 \bar{D}_0 + F_1 \bar{D}_1. \qquad (3.4.2)$$

We can express this optimization problem as follows:

$$\max_v \inf_{\mu,x,u} \left\{ x(0)^T P_\beta^{-1} x(0) + \int_0^T \mu^T \Gamma_{\beta,\gamma} \mu \, dt \right\} \qquad (3.4.3)$$

subject to

$$\dot{x} = Ax + \mathcal{B}\mu + \bar{B}u, \qquad (3.4.4a)$$

$$0 = Cx + \mathcal{D}\mu + \bar{D}u, \qquad (3.4.4b)$$

where $\mu = (\nu, v)$,

$$\mathcal{B} = \begin{pmatrix} M & B \end{pmatrix}, \qquad (3.4.5)$$

$$\mathcal{D} = \begin{pmatrix} N & D \end{pmatrix}, \qquad (3.4.6)$$

and

$$\Gamma_{\beta,\gamma} = \begin{pmatrix} J_\beta & 0 \\ 0 & -\lambda I \end{pmatrix}. \tag{3.4.7}$$

Note that if $\bar{B} = 0$ and $\bar{D} = 0$ and consequently u is no longer present, we already know how to solve the problem. So in order to solve this problem, we propose a procedure for removing u from the constraints.

The constraints (3.4.4) can be expressed as follows:

$$\mathcal{E}\dot{x} = \mathcal{F}x + \mathcal{G}\mu + \mathcal{Z}u \tag{3.4.8}$$

where

$$\mathcal{E} = \begin{pmatrix} I \\ 0 \end{pmatrix}, \quad \mathcal{F} = \begin{pmatrix} A \\ C \end{pmatrix}, \quad \mathcal{G} = \begin{pmatrix} B \\ D \end{pmatrix}, \quad \mathcal{Z} = \begin{pmatrix} \bar{B} \\ \bar{D} \end{pmatrix}. \tag{3.4.9}$$

Note that $\mathcal{E}$ has full column rank.

The signal v must be proper no matter what u turns out to be. Thus in terms of computing v, we must view u as an unknown arbitrary function. We can eliminate u from the constraints (3.4.8) by multiplying on the left by $\mathcal{Z}^\perp$, a highest-rank left annihilator of $\mathcal{Z}$. This yields the constraint

$$\mathcal{E}_0\dot{x} = \mathcal{F}_0 x + \mathcal{G}_0\mu \tag{3.4.10}$$

with $\mathcal{E}_0 = \mathcal{Z}^\perp\mathcal{E}, \mathcal{F}_0 = \mathcal{Z}^\perp\mathcal{F}, \mathcal{G}_0 = \mathcal{Z}^\perp\mathcal{G}$ which can replace the constraints (3.4.4) without affecting the optimization problem.

If $\mathcal{E}_0$ is full column rank, we can put (3.4.10) directly into the form (3.4.4) but with $\bar{B} = 0$, $\bar{D} = 0$, and we are done. If not, we can find an invertible matrix Ω_0 such that

$$\mathcal{E}_0\Omega_0 = \begin{pmatrix} \mathcal{E}_{00} & 0 \end{pmatrix} \tag{3.4.11}$$

with $\mathcal{E}_{00}$ being full column rank. By performing the change of coordinates

$$\begin{pmatrix} x_{00} \\ x_{01} \end{pmatrix} = \Omega_0^{-1}x \tag{3.4.12}$$

we can rewrite (3.4.10) as follows:

$$\mathcal{E}_{00}\dot{x}_{00} = \mathcal{F}_{00}x_{00} + \mathcal{G}_0\mu - \mathcal{E}_{01}x_{01}. \tag{3.4.13}$$

Note that $\mathcal{E}_{00}$ is full column rank. The problem is now the term $\mathcal{E}_{01}x_{01}$. We can eliminate this by multiplying (3.4.13) on the left with a highest-rank left annihilator of $\mathcal{E}_{01}$. This can be done because x (and thus x_{01}) does not appear explicitly in the integral part of the cost function (3.4.3). In fact x_{01} is free just as u was. This multiplication yields

$$\mathcal{E}_1\dot{x}_{00} = \mathcal{F}_1 x_{00} + \mathcal{G}_1\mu. \tag{3.4.14}$$

Moreover, note that after the coordinate transformation the cost function can be written as follows:

$$\max_v \inf_{\mu,x,u} \left\{ \begin{pmatrix} x_{00}(0) \\ x_{01}(0) \end{pmatrix}^T \begin{pmatrix} \Psi_0 & \Psi_1 \\ \Psi_1^T & \Psi_2 \end{pmatrix} \begin{pmatrix} x_{00}(0) \\ x_{01}(0) \end{pmatrix} + \int_0^T \mu^T\Gamma_{\beta,\gamma}\mu \, dt \right\}, \tag{3.4.15}$$

where

$$\begin{pmatrix} \Psi_0 & \Psi_1 \\ \Psi_1^T & \Psi_2 \end{pmatrix} = \Omega_0^T P_\beta^{-1} \Omega_0. \tag{3.4.16}$$

Since $x_{01}(0)$ is not in any way constrained (we are allowing for discontinuity in x, or equivalently impulsive u's), the minimization over $x_{01}(0)$ can be carried out directly. It is easy to prove the following lemma.

Lemma 3.4.1

$$\min_{x_{01}(0)} \begin{pmatrix} x_{00}(0) \\ x_{01}(0) \end{pmatrix}^T \begin{pmatrix} \Psi_0 & \Psi_1 \\ \Psi_1^T & \Psi_2 \end{pmatrix} \begin{pmatrix} x_{00}(0) \\ x_{01}(0) \end{pmatrix} = x_{00}(0)^T P_{\beta,0}^{-1} x_{00}(0), \tag{3.4.17}$$

where

$$P_{\beta,0}^{-1} = \Psi_0 - \Psi_1^T \Psi_2^{-1} \Psi_1. \tag{3.4.18}$$

So now the optimization problem can be expressed as follows:

$$\max_v \inf_{\mu, x_{00}, u} \left\{ x_{00}(0)^T P_{\beta,0}^{-1} x_{00}(0) + \int_0^T \mu^T \Gamma_{\beta,\gamma} \mu \, dt \right\} \tag{3.4.19}$$

subject to (3.4.14). But this is similar to our previous optimization problem (3.4.3) subject to (3.4.10). Thus we can repeat the procedure: if $\mathcal{E}_1$ is full column rank, we can put (3.4.14) directly into the form (3.4.4). If $\bar{B} = 0$ and $\bar{D} = 0$, then we are done. If not, then we can find an invertible matrix Ω_1 such that

$$\mathcal{E}_1 \Omega_1 = \begin{pmatrix} \mathcal{E}_{10} & 0 \end{pmatrix} \tag{3.4.20}$$

with $\mathcal{E}_{10}$ being full column rank. By performing the change of coordinates

$$\begin{pmatrix} x_{10} \\ x_{11} \end{pmatrix} = \Omega_1^{-1} x \tag{3.4.21}$$

we can rewrite (3.4.14) as

$$\mathcal{E}_{10} \dot{x}_{10} = \mathcal{F}_{10} x_{10} + \mathcal{G}_1 \mu - \mathcal{E}_{11} x_{11} \tag{3.4.22}$$

and continue as in the previous case.

Note that the size of x_{k0} is a strictly decreasing function of k so the algorithm stops after a finite number of steps. This recursive procedure can be implemented efficiently in a few lines of code in Scilab or Matlab.

This reduction can be carried out for any system of the form (3.4.1). At the end of the reduction, if condition 1 of theorem 3.3.1 is satisfied by the reduced system, then we can proceed as before and construct that auxiliary signal.

The problem considered in this section is closely related to the singular linear quadratic (LQ) problem and, in particular, the work of [82]. Here, however, the problem is simpler because we do not need to reconstruct the "control" u.

3.5 MORE GENERAL COST FUNCTIONS

Up to now, we have considered only the case where our objective has been to minimize the energy of the auxiliary signal. This is reasonable because it usually implies that the system being tested is not affected too much by the test. In some cases, however, we need to be more specific about the way we want the system to be protected from the effects of the auxiliary signal. For example, we may want to make sure that at the end of the test period, if the system has not failed, then the state of the system will be close to zero. Or we may want to have a specific component of the system be less affected by the auxiliary signal while another component is allowed to be heavily affected. To be able to take into account this kind of consideration, we have to generalize the optimization problem (3.3.13) as follows:

$$\min_v \xi(T)^T W \xi(T) + \int_0^T |v|^2 + \xi^T U \xi \, dt \quad \text{subject to} \quad \max_{\substack{\beta \in [0,1] \\ s \in [0,T]}} \phi_\beta(v, s) \geq 1,$$

$$(3.5.1a)$$

where W and U are positive semidefinite matrices with appropriate sizes (U may depend on t) and ξ is a vector function of time satisfying

$$\dot{\xi} = F\xi + Gv, \quad \xi(0) = 0. \tag{3.5.1b}$$

Here F and G are matrices, with appropriate dimensions, chosen by design considerations.

In many applications, F and G are, respectively, A_0 and B_0 so that (3.5.1b) represents the normal behavior of the system, without the uncertainties. This means in particular that $\xi(t)$ is an a priori estimate of $x_0(t)$, and thus it amounts to penalizing the perturbation of the system during the test period (assuming no failure has occurred). It is also possible to consider ξ to contain both an estimate of x_0 and x_1 in order to reduce the effect of the auxiliary signal on the behavior of the system whether or not a failure has occurred. And, of course, there are many other scenarios which can be envisaged and modeled by (3.5.1b),(3.5.1a).

3.5.1 Computation of the Separability Index

The solution of the optimization problem (3.5.1a) can be derived by a straightforward modification of what was done in section 3.3.1. In particular, we define

$$\lambda_{\beta,s} = \max_{v \neq 0} \frac{\phi_\beta(v, s)}{\xi(s)^T W \xi(s) + \int_0^s |v|^2 + \xi^T U \xi \, dt} \tag{3.5.2}$$

and λ_β is defined as before by

$$\lambda_\beta = \max_{s \leq T} \lambda_{\beta,s} \tag{3.5.3}$$

so that we end up having to solve the following problem:

$$\max_v \inf_{v,x} x(0)^T P_\beta^{-1} x(0) - \lambda \xi(T)^T W \xi(T) + \int_0^s v^T J_\beta v - \lambda(\xi^T U \xi + |v|^2) \, dt$$

$$(3.5.4)$$

subject to the constraints (3.5.1b) and (3.3.7). We can rewrite these constraints as follows:

$$\dot{\mu} = \begin{pmatrix} A & 0 \\ 0 & F \end{pmatrix} \mu + \begin{pmatrix} M & B \\ 0 & G \end{pmatrix} \begin{pmatrix} \nu, \\ v \end{pmatrix} \tag{3.5.5}$$

$$0 = \begin{pmatrix} C & 0 \end{pmatrix} \mu + \begin{pmatrix} N & D \end{pmatrix} \begin{pmatrix} \nu \\ v \end{pmatrix}, \tag{3.5.6}$$

$$0 = \begin{pmatrix} 0 & I \end{pmatrix} \mu(0), \tag{3.5.7}$$

where

$$\mu = \begin{pmatrix} x \\ \xi \end{pmatrix}. \tag{3.5.8}$$

The cost in (3.5.4) then becomes

$$\max_{v} \min_{\nu,\mu} \left\{ \mu(0)^T \begin{pmatrix} P_\beta^{-1} & 0 \\ 0 & 0 \end{pmatrix} \mu(0) + \mu(s)^T \begin{pmatrix} 0 & 0 \\ 0 & -\lambda W \end{pmatrix} \mu(s) \right.$$

$$\left. + \int_0^s \nu^T J_\beta \nu - \lambda |v|^2 - \mu^T \mathcal{Q}_\lambda \mu \, dt \right\}, \tag{3.5.9}$$

where

$$\mathcal{Q}_\lambda = \begin{pmatrix} 0 & 0 \\ 0 & \lambda U \end{pmatrix}. \tag{3.5.10}$$

By defining the corresponding Lagrangian as usual and setting its first variation with respect to v, ν, and μ in (3.5.9) to zero, we obtain the following two-point boundary value system after some straightforward algebra:

$$\frac{d}{dt} \begin{pmatrix} \mu \\ \zeta \end{pmatrix} = \begin{pmatrix} \bar\Omega_{11} & \bar\Omega_{12} \\ \bar\Omega_{21} & \bar\Omega_{22} \end{pmatrix} \begin{pmatrix} \mu \\ \zeta \end{pmatrix} \tag{3.5.11a}$$

with boundary conditions

$$V_0 \begin{pmatrix} \mu(0) \\ \zeta(0) \end{pmatrix} + V_s \begin{pmatrix} \mu(s) \\ \zeta(s) \end{pmatrix} = 0. \tag{3.5.11b}$$

Here

$$V_0 = \begin{pmatrix} I & \begin{pmatrix} -P_\beta & 0 \\ 0 & 0 \end{pmatrix} \\ 0 & 0 \end{pmatrix}, \tag{3.5.12}$$

$$V_s = \begin{pmatrix} 0 & 0 \\ \begin{pmatrix} 0 & 0 \\ 0 & -\lambda W \end{pmatrix} & I \end{pmatrix}, \tag{3.5.13}$$

and

$$\bar\Omega_{11} = -\bar\Omega_{22}^T = \bar A - \bar S_{\lambda,\beta} \bar R_{\lambda,\beta}^{-1} \bar C, \tag{3.5.14}$$

$$\bar\Omega_{12} = \bar Q_{\lambda,\beta} - \bar S_{\lambda,\beta} \bar R_{\lambda,\beta}^{-1} \bar S_{\lambda,\beta}^T, \tag{3.5.15}$$

$$\bar\Omega_{21} = \bar C^T R_{\lambda,\beta}^{-1} \bar C + \mathcal{Q}_\lambda, \tag{3.5.16}$$

where

$$\begin{pmatrix} \bar Q_{\lambda,\beta} & \bar S_{\lambda,\beta} \\ \bar S_{\lambda,\beta}^T & \bar R_{\lambda,\beta} \end{pmatrix} = \begin{pmatrix} \bar B \\ \bar D \end{pmatrix} \Gamma_{\lambda,\beta}^{-1} \begin{pmatrix} \bar B \\ \bar D \end{pmatrix}^T, \tag{3.5.17}$$

$$\Gamma_{\lambda,\beta} = \begin{pmatrix} J_\beta & 0 \\ 0 & -\lambda I \end{pmatrix}, \tag{3.5.18}$$

$$\bar A = \begin{pmatrix} A & 0 \\ 0 & F \end{pmatrix}, \quad \bar B = \begin{pmatrix} M & B \\ 0 & G \end{pmatrix}, \tag{3.5.19}$$

$$\bar C = \begin{pmatrix} C & 0 \end{pmatrix}, \quad \bar D = \begin{pmatrix} N & D \end{pmatrix}. \tag{3.5.20}$$

Then the optimal v and ν for (3.5.9) satisfy

$$\begin{pmatrix} \nu^* \\ v^* \end{pmatrix} = \alpha \Gamma_{\lambda,\beta}^{-1} \left(\bar{D}^T \bar{R}_{\lambda,\beta}^{-1} \bar{C} \quad \bar{D}^T \bar{R}_{\lambda,\beta}^{-1} \bar{S}_{\lambda,\beta}^T - \bar{B}^T \right) \begin{pmatrix} \mu \\ \xi \end{pmatrix}, \tag{3.5.21}$$

where α is an arbitrary scalar.

We need to compute first λ_β, which is the largest value of λ for which the two-point boundary value system (3.5.11) is not well posed for some $s \in [0, T]$. The following result characterizes the well-posedness of our boundary value system.

Lemma 3.5.1 *The two-point boundary value system*

$$\dot{x} = Hx, \tag{3.5.22a}$$

$$0 = V_i x(0) + V_f x(s) \tag{3.5.22b}$$

is well posed if and only if $V_i + V_f \Phi(s)$ is invertible, where

$$\dot{\Phi} = H\Phi, \quad \Phi(0) = I. \tag{3.5.23}$$

The proof is straightforward and uses the fact that

$$x(s) = \Phi(s)x(0). \tag{3.5.24}$$

It is rather simple to apply this result to our problem by taking

$$H = \Omega = \begin{pmatrix} \bar{\Omega}_{11} & \bar{\Omega}_{12} \\ \bar{\Omega}_{21} & \bar{\Omega}_{22} \end{pmatrix} \tag{3.5.25}$$

and $V_i = V_0$ and $V_f = V_s$. The problem is that the computation of Φ based on (3.5.23) is in general not practical. Since H is a Hamiltonian matrix, it generally has eigenvalues of both negative and positive real parts. Thus numerical integration of (3.5.23) could be numerically unstable. So this approach can be considered only when the test period is very short.

3.5.1.1 The Riccati Approach

To compute λ_β, we can proceed as before using a Riccati equation but the situation is more complex, in particular due to the presence of the final cost term on $\mu(s)$.

Consider the Riccati equation

$$\dot{P} = (\bar{A} - \bar{S}_{\lambda,\beta} \bar{R}_{\lambda,\beta}^{-1} \bar{C})P + P(\bar{A} - \bar{S}_{\lambda,\beta} \bar{R}_{\lambda,\beta}^{-1} \bar{C})^T$$
$$- P(\bar{C}^T \bar{R}_{\lambda,\beta}^{-1} \bar{C} + Q_\lambda)P + \bar{Q}_{\lambda,\beta} - \bar{S}_{\lambda,\beta} \bar{R}_{\lambda,\beta}^{-1} \bar{S}_{\lambda,\beta}^T, \quad P(0) = \begin{pmatrix} P_\beta & 0 \\ 0 & 0 \end{pmatrix}. \tag{3.5.26}$$

By applying the change of coordinate

$$x = \begin{pmatrix} I & P \\ 0 & I \end{pmatrix} \begin{pmatrix} \mu \\ \varsigma \end{pmatrix} \tag{3.5.27}$$

the two-point boundary value system (3.5.11) becomes

$$\dot{x} = \begin{pmatrix} \bar{\Omega}_{11} - P\bar{\Omega}_{21} & 0 \\ \bar{\Omega}_{21} & (\bar{\Omega}_{11} - P\bar{\Omega}_{21})^T \end{pmatrix} x \tag{3.5.28a}$$

with boundary conditions

$$V_i x(0) + V_f x(s) = 0, \tag{3.5.28b}$$

where

$$V_i = \begin{pmatrix} I & 0 \\ 0 & 0 \end{pmatrix}, \tag{3.5.29a}$$

$$V_f = \begin{pmatrix} \begin{pmatrix} 0 & 0 \\ 0 & \lambda W \end{pmatrix} & I - \begin{pmatrix} 0 & 0 \\ 0 & \lambda W \end{pmatrix} P \end{pmatrix}. \tag{3.5.29b}$$

By applying lemma 3.5.1 to this two-point boundary value system and using the block triangular nature of its dynamics, we see that the two-point boundary value system is not well posed if

$$\det \left(I - \begin{pmatrix} 0 & 0 \\ 0 & \lambda W \end{pmatrix} P \right) = 0. \tag{3.5.30}$$

So the well-posedness test can be done by integrating the Riccati equation (3.5.26) and checking for condition (3.5.30). This type of test can, in general, be implemented using any standard ordinary differential solver with a root finder option. Here, however, there is an additional difficulty in that the Riccati equation (3.5.26) must be integrated through singularities. In practice, this can be done by switching back and forth from the Riccati equation (3.5.26) to a Riccati equation in terms of the inverse of P denoted $\bar{P}$:

$$-\dot{\bar{P}} = \bar{P}(\bar{A} - \bar{S}_{\lambda,\beta}\bar{R}_{\lambda,\beta}^{-1}\bar{C}) + (\bar{A} - \bar{S}_{\lambda,\beta}\bar{R}_{\lambda,\beta}^{-1}\bar{C})^T \bar{P}$$
$$- (\bar{C}^T \bar{R}_{\lambda,\beta}^{-1}\bar{C} + \mathcal{Q}_\lambda) + \bar{P}(\bar{\mathcal{Q}}_{\lambda,\beta} - \bar{S}_{\lambda,\beta}\bar{R}_{\lambda,\beta}^{-1}\bar{S}_{\lambda,\beta}^T)\bar{P}. \tag{3.5.31}$$

An algorithm for the computation of the λ_β can now be constructed by noting that $\lambda > \lambda_\beta$ if and only if the surface (3.5.30) is not crossed during the interval $[0, T]$. A Scilab code implementing this procedure is given in section 6.2 of chapter 6.

This Riccati-based procedure works even if the system is time varying. But when the system is not time varying (which is the case for most applications), a simpler approach can be used, leading to a more efficient code.

3.5.1.2 The Block Diagonalization Approach

Consider the two-point boundary value system (3.5.22). When H is constant, it is possible to construct a simple and numerically efficient test of its well-posedness by block diagonalizing H as follows:

$$SHS^{-1} = \begin{pmatrix} A_f & 0 \\ 0 & -A_b \end{pmatrix}, \tag{3.5.32}$$

where A_f and A_b do not have any eigenvalues with strictly positive real parts. The well-posedness of the two-point boundary value system is equivalent to the invertibility of $V_i + V_f \Phi(s)$, which can be tested by examining the invertibility of

$$(V_i + V_f \Phi(s))S^{-1} = V_i S^{-1} + V_f S^{-1} \begin{pmatrix} \exp(A_f s) & 0 \\ 0 & \exp(-A_b s) \end{pmatrix}. \tag{3.5.33}$$

The problem is that $\exp(-A_b s)$ diverges. But since all that we care about is the invertibility of the matrix in (3.5.33), we can eliminate the term $\exp(-A_b s)$ by postmultiplying (3.5.33) with the invertible matrix

$$\begin{pmatrix} I & 0 \\ 0 & \exp(A_b s) \end{pmatrix},$$

obtaining

$$\Psi(s) \triangleq V_i S^{-1} \begin{pmatrix} I & 0 \\ 0 & \exp(A_b s) \end{pmatrix} + V_f S^{-1} \begin{pmatrix} \exp(A_f s) & 0 \\ 0 & I \end{pmatrix}. \qquad (3.5.34)$$

We can now construct a simple procedure for testing whether or not a given λ is larger than λ_β.

Lemma 3.5.2 *Suppose the two conditions of theorem 3.3.1 are satisfied. Then $\lambda > \lambda_\beta$ if and only if $\Psi(s)$ is invertible for all $s \in [0, T]$.*

A λ-iteration scheme can now be implemented using any standard ordinary differential equation solver with a root finder option. In particular, we have to solve

$$\dot{\Psi}_f = A_f \Psi_f, \quad \Psi_f(0) = I, \qquad (3.5.35a)$$

$$\dot{\Psi}_b = A_b \Psi_b, \quad \Psi_b(0) = I, \qquad (3.5.35b)$$

and test to see if the surface

$$0 = \det \left(V_i S^{-1} \begin{pmatrix} I & 0 \\ 0 & \Psi_b \end{pmatrix} + V_f S^{-1} \begin{pmatrix} \Psi_f & 0 \\ 0 & I \end{pmatrix} \right) \qquad (3.5.36)$$

is crossed. Then λ_β is obtained by taking the infimum over the set of λ's for which the above system can be solved over the interval $[0, T]$ without any surface crossing. A Scilab program implementing this procedure can be found in section 6.3 of chapter 6.

Once we have a procedure for computing λ_β, λ^* and β^* are obtained as before by

$$\lambda^* = \max_{\beta \in \mathcal{B}} \lambda_\beta, \qquad (3.5.37)$$

giving the separability index

$$\gamma^* = \sqrt{\lambda^*}. \qquad (3.5.38)$$

Note that in both approaches, for $\lambda = \lambda_\beta$, the surface crossing does not necessarily occur at $t = T$; it may happen inside the interval $[0, T]$, say at T^*. This simply means that the optimal proper auxiliary signal can be defined over the interval $[0, T^*]$, nothing can be gained by increasing the length of the test period, and so we can let $T = T^*$.

To understand why in some cases $T^* < T$, see figure 3.5.1, where we have illustrated $\lambda_{\beta,s}$ and $\lambda_\beta(s) = \max_{t \le s} \lambda_{\beta,t}$ in a typical situation. Note that $\lambda_{\beta,s}$ is not necessarily an increasing function of s, but $\lambda_\beta(s)$ is.

Suppose $T = T_1$, in this case, we have $T^* = T = T_1$. But, if $T = T_2$, it is clear that as we increase the length of the interval, we find a singularity at $T^* < T$ as illustrated on the figure. Also note that as s goes to infinity we have the optimal λ converging to λ_β^*, which can be interpreted as the best that could be done if there were no limit on the length of the test period.

We assume for the rest of this chapter that $T^* = T$.

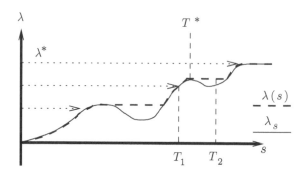

Figure 3.5.1   Study of the well-posedness of the two-point boundary
value system as a function of its interval length.

3.5.2 Construction of an Optimal Auxiliary Signal

When $\lambda = \lambda^*$ and $\beta = \beta^*$, the two-point boundary value system (3.5.11) is not well posed, i.e., it has a nonzero solution. It is this solution that allows us to compute the optimal proper auxiliary signal from (3.5.21).

In the Riccati approach, we can construct the solution of the two-point boundary value system using the solution of the Riccati equation as we did in section 3.3.3. The existence of singularities in this case makes the numerical implementation a bit more difficult.

The block diagonalization approach, on the other hand, results in simple and numerically reliable code. This approach, of course, applies only if the system is time invariant.

Lemma 3.5.3 *The initial condition $x(0)$ and function $x(s)$ are a solution of the two-point boundary value system (3.5.22) if and only if*

$$\begin{pmatrix} -\Phi(s) & I \\ V_i & V_f \end{pmatrix} \begin{pmatrix} x(0) \\ x(s) \end{pmatrix} = 0. \tag{3.5.39}$$

Equation (3.5.39) when $s = T$, after the coordinate transformation S, can be expressed as follows:

$$\left(\begin{pmatrix} -\Psi_f(T) & 0 \\ 0 & I \\ V_i S^{-1} \end{pmatrix} \begin{pmatrix} I & 0 \\ 0 & -\Psi_b(T) \\ V_f S^{-1} \end{pmatrix} \right) \begin{pmatrix} x_f(0) \\ x_b(0) \\ x_f(T) \\ x_b(T) \end{pmatrix} = 0, \tag{3.5.40}$$

where

$$\begin{pmatrix} x_f(0) \\ x_b(0) \\ x_f(T) \\ x_b(T) \end{pmatrix} = \begin{pmatrix} Sx(0) \\ Sx(T) \end{pmatrix}. \tag{3.5.41}$$

Thus by computing a vector in the nullspace of the matrix

$$\left(\begin{pmatrix} -\Psi_f(T) & 0 \\ 0 & I \\ V_i S^{-1} \end{pmatrix} \begin{pmatrix} I & 0 \\ 0 & -\Psi_b(T) \\ V_f S^{-1} \end{pmatrix} \right),$$

we can find consistent values of $x_f(0)$ and $x_b(T)$. But x_f and x_b satisfy

$$\dot{x}_f = A_f x_f, \tag{3.5.42a}$$

$$\dot{x}_b = -A_b x_b \tag{3.5.42b}$$

with A_f and A_b that do not have eigenvalues with positive real parts. This means that x_f and x_b can be computed from (3.5.42a) and (3.5.42b), respectively, by forward and backward integration. Finally, the solution of the boundary value problem is obtained from

$$x = S^{-1} \begin{pmatrix} x_f \\ x_b \end{pmatrix}. \tag{3.5.43}$$

In our case,

$$x = \begin{pmatrix} \mu \\ \xi \end{pmatrix}, \tag{3.5.44}$$

so the optimal auxiliary signal can now be computed from (3.5.21) by choosing α such that $\|v^*\| = 1/\gamma^*$.

3.6 DISCRETE-TIME CASE

The discrete-time theory can be developed along the same lines as the continuous-time theory. We suppose that the models of the normal and failed systems can be expressed as follows:

$$x_i(k+1) = A_i x_i(k) + B_i v(k) + \bar{B}_i u(k) + M_i \nu_i(k), \tag{3.6.1a}$$

$$E_i y(k) = C_i x_i(k) + D_i v(k) + \bar{D}_i u(k) + N_i \nu_i(k), \tag{3.6.1b}$$

where all the matrices may depend on k. The model with $i = 0$ corresponds to the normal system and the model with $i = 1$ is the failed system. The E_i's have full column rank and the N_i's have full row rank.

The constraint (or noise measure) on the initial condition and uncertainties (2.3.24) is

$$S_i(v, k) = x_i(0)^T P_{i0}^{-1} x_i(0) + \sum_{i=0}^{k} \nu_i(i)^T J_i \nu_i(i) < 1, \ \forall \, k \in [0, N-1], \tag{3.6.2}$$

where the J_i's are signature matrices. As discussed in chapter 2, the bound (3.6.2) allows for both additive and model uncertainty.

We define the function σ as we did in the continuous-time case:

$$\sigma(v, k) = \inf_{\substack{\nu_0, \nu_1, u, y \\ x_0, x_1}} \max(S_0(v, k), S_1(v, k)). \tag{3.6.3}$$

Then, since the N_i's have full row rank, the nonexistence of a solution to (3.6.1a), (3.6.1b), and (3.6.2) is equivalent to

$$\sigma(v, k) \geq 1. \tag{3.6.4}$$

We can express $\sigma(v, k)$ as follows:

$$\sigma(v, k) = \max_{\beta \in [0,1]} \phi_\beta(v, k), \tag{3.6.5a}$$

where

$$\phi_\beta(v,k) = \inf_{\substack{\nu_0,\nu_1,u,y \\ x_0,x_1}} \beta S_0(v,k) + (1-\beta)S_1(v,k) \tag{3.6.5b}$$

subject to (3.6.1a),(3.6.1b), $i = 0,1$.

Using the notation (3.3.5) and assuming there is no u (the case with u can be transformed to a case without u as we did in the continuous-time case), we can reformulate the problem (3.6.5) as follows:

$$\phi_\beta(v,k) = \inf_{\nu,x} x(0)^T P_\beta^{-1} x(0) + \sum_{i=0}^{k} \nu(i)^T J_\beta \nu(i) \tag{3.6.6}$$

subject to

$$x(i+1) = Ax(i) + Bv(i) + M\nu(i), \tag{3.6.7a}$$
$$0 = Cx(i) + Dv(i) + N\nu(i). \tag{3.6.7b}$$

With arguments similar to those used in the continuous-time case, we can show that the construction of the optimal proper auxiliary signal amounts to solving

$$\lambda = \max_{v,\beta,k} \frac{\phi_\beta(v,k)}{q_k(v_0,\dots,v_k)} \tag{3.6.8}$$

where the q_k are quadratic positive functions corresponding to the cost on v.

The discrete-time problem we are considering is defined over a finite interval. Thus it is straightforward to convert this maximization-minimization optimization problem into a static problem and solve it using the techniques discussed earlier in this chapter. Note, however, that the size of the matrices involved could be very large if the size of the state $x(i)$ is large. Even though we have no constraints on the computation time (all the computation is done off-line), the amount of memory needed for the computation can be prohibitive.

Another approach would be to develop a method similar to the method derived for the continuous-time case taking full advantage of the recursive nature of the problem. Here we present an intermediate solution, which retains some of the simplicity of a static approach yet avoids in most cases having to construct huge matrices. This method was first presented in [66].

The idea is to use recursive formulas to solve the minimization problem (3.6.6), but to solve the maximization problem over v as a static problem. This method can be efficiently implemented as long as the length of the interval times the size of $v(i)$ is not too large. This is often true since $v(i)$ usually has much lower dimension than $x(i)$.

Theorem 3.6.1 *Suppose*

1. Either $\begin{pmatrix} A & M \\ C & N \end{pmatrix}$ *is invertible or*

$$\begin{pmatrix} A & M \\ C & N \end{pmatrix}_\perp^T \begin{pmatrix} P_j^{-1} & 0 \\ 0 & J_\beta \end{pmatrix} \begin{pmatrix} A & M \\ C & N \end{pmatrix}_\perp > 0, \ \forall\, j; \tag{3.6.9}$$

2. *The Riccati equation*

$$P_{j+1} = (AP_jA^T + MJ_\beta^{-1}M^T)$$
$$-(AP_jC^T + MJ_\beta^{-1}N^T)(CP_jC^T + NJ_\beta^{-1}N^T)^{-1}(AP_jC^T + MJ_\beta^{-1}N^T)^T$$
$$(3.6.10)$$

has positive solution P_j on $[1, k]$.

Then the solution to the optimization problem (3.6.6) is given by

$$\phi_\beta(v, k) = v_k^T \Omega_{k,\beta} v_k, \tag{3.6.11}$$

where

$$v_k = \begin{pmatrix} v(0) \\ \vdots \\ v(k) \end{pmatrix} \tag{3.6.12}$$

and $\Omega_{k,\beta}$ is obtained from the following recursion:

$$\Omega_{0,\beta} = D^T \Delta_0^{-1} D, \tag{3.6.13a}$$

$$\Omega_{j+1,\beta} = \begin{pmatrix} \Omega_{j,\beta} & 0 \\ 0 & 0 \end{pmatrix} + \begin{pmatrix} Q_{j+1,\beta}C^T \\ D^T \end{pmatrix} \Delta_{j+1,\beta}^{-1} \begin{pmatrix} CQ_{j+1,\beta}^T & D \end{pmatrix}. \tag{3.6.13b}$$

Here the Q_j satisfy the recursion

$$Q_{0,\beta} = [] \text{ (empty matrix)}; \tag{3.6.14a}$$

$$Q_{j+1,\beta} = \begin{pmatrix} (A - (AP_jC^T + MJ_\beta^{-1}N^T\Delta_{j,\beta}^{-1})C)Q_j \\ B - (AP_jC^T + MJ_\beta^{-1}N^T)\Delta_{j,\beta-1}D \end{pmatrix}, \tag{3.6.14b}$$

where

$$\Delta_{j,\beta} = CP_jC^T + NJ_\beta^{-1}N^T. \tag{3.6.15}$$

This result is a straightforward consequence of theorem 2.6.4 and its corollary.

Since the cost on v, denoted by q_k, is quadratic and positive in v_k, we can express it as

$$q_k(v_0, \dots, v_k) = v_k^T V_k v_k, \tag{3.6.16}$$

where V_k is a positive definite matrix. The procedure for the computation of the optimal auxiliary signal can now be implemented as follows. Compute $\Omega_{k,\beta}$ for various values of k and β where the assumptions of theorem 3.6.1 hold. Find

$$\lambda = \max_{k,\beta} \left(\text{largest eigenvalue of } \Omega_{k,\beta} V_k^{-1} \right). \tag{3.6.17}$$

This determines the optimal interval length k^* and an optimal β denoted β^*. Then use λ to construct the optimal v_{k^*} by solving the generalized eigenvalue problem

$$(\lambda V_{k^*} - \Omega_{k^*,\beta^*})v_{k^*} = 0. \tag{3.6.18}$$

Note that v_{k^*} should be normalized as in the continuous-time case.

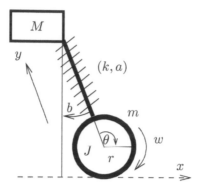

Figure 3.7.1 Suspension example.

3.7 SUSPENSION EXAMPLE

In this section we introduce the particular example of a vehicle suspension. The model is somewhat idealized but serves to illustrate the application of the method presented in this chapter and will be returned to later in the book.

3.7.1 System Model

The suspension is shown in figure 3.7.1. It is modeled as a mass M attached to a rolling wheel through a spring mass system. The control is a torque applied to the wheel. Friction in the wheel and axle is ignored. The model is

$$M\ddot{y} - Mr\sin(b)\ddot{\theta} + a\dot{y} + ky = 0, \qquad (3.7.1a)$$

$$-Mr\sin(b)\ddot{y} + ((m+M)r^2 + J)\ddot{\theta} = w, \qquad (3.7.1b)$$

where M is the mass of the suspended body (car, motorcycle, etc.), a and k are suspension parameters, J, m, and r are wheel parameters (rotational inertia, mass, and radius), b is the angle the suspension makes with the vertical (assumed constant), w is the torque variation on the wheel, θ measures the rotation of the wheel, and finally y is the variation in the length of the suspension from its rest length. If wheel friction is included in (3.7.1), then a $\dot{\theta}$ term is added to (3.7.1b). Note that $y = 0$, $\theta = \theta_0 + \alpha t$ is a solution of (3.7.1) when $w = 0$. This corresponds to traveling along a level road at a constant speed with no disturbances. We will replace θ in (3.7.1) by $\theta - \theta_0 - \alpha t$ so that θ is now the deviation from this reference angle and disturbances in the initial conditions are disturbances from $y = 0$, $\theta = \theta_0 + \alpha t$. If wheel friction is included, then w becomes the change in torque from w_o, where w_o is the constant torque needed to maintain the speed α.

The input to the system is w and the outputs are (noisy) measurements of the suspension length y and the rotational rate $\dot{\theta}$ so that the output equations, without the measurement noise, are

$$y_1 = y, \qquad (3.7.2a)$$

$$y_2 = \dot{\theta}. \qquad (3.7.2b)$$

3.7.2 Additive Noise Formulation

The system (3.7.1) is already linear and can be put into the standard first-order form by state augmentation, for example, by defining the state as $y, \dot{y}, \theta$, and $\dot{\theta}$.

We will first consider the case of additive noise. Note that, even if both equations in (3.7.1) have additive noise, only two of the four equations in the first-order version of the model would have noise since there would be no noise in the augmentation equations $x_1 = \dot{y}, x_3 = \dot{\theta}$. This situation is common in applications and illustrates why we never assumed the matrix M was full row rank. Here we will assume that the noise in the suspension itself is negligible and that there is a noise variable added to (3.7.1b). For these equations one could think of it as noise (in force units) arising from irregularities in the road surface. In addition, we will assume an additive noise on each of the output channels to represent measurement noise. The failure to be modeled is a complete failure of the sensor measuring y. That is, $y_1 = 0$.

Let

$$\mathcal{P} = \begin{pmatrix} M & -Mr\sin(b) \\ -Mr\sin(b) & (m+M)r^2 + J \end{pmatrix}, \quad \mathcal{Q} = \begin{pmatrix} a & 0 \\ 0 & 0 \end{pmatrix}, \quad \mathcal{R} = \begin{pmatrix} k & 0 \\ 0 & 0 \end{pmatrix}. \quad (3.7.3)$$

Then it is straightforward to show that the normal system can be modeled as in (3.3.1a), where

$$A_0 = \begin{pmatrix} 0 & I \\ -\mathcal{P}^{-1}\mathcal{R} & -\mathcal{P}^{-1}\mathcal{Q} \end{pmatrix}, \quad (3.7.4)$$

$$B_0 = \begin{pmatrix} 0 \\ \mathcal{P}^{-1}\begin{pmatrix} 0 \\ 1 \end{pmatrix} \end{pmatrix}, \quad (3.7.5)$$

$$C_0 = \begin{pmatrix} 1 & 0 & 0 & 0 \\ 0 & 0 & 0 & 1 \end{pmatrix}, \quad D_0 = 0. \quad (3.7.6)$$

The system matrices for the additive noises are defined to be

$$M_0 = \begin{pmatrix} 0 & 0 & 0 \\ 0 & 0 & 0 \\ 0 & 0 & 0 \\ 0.001 & 0 & 0 \end{pmatrix}, \quad N_0 = \begin{pmatrix} 0 & 0.01 & 0 \\ 0 & 0 & 0.01 \end{pmatrix}. \quad (3.7.7)$$

The failed system is identical to the normal system except for the C matrix, i.e., $A_1 = A_0, B_1 = B_0, D_1 = D_0, M_1 = M_0, N_1 = N_0$, and

$$C_1 = \begin{pmatrix} 0 & 0 & 0 & 0 \\ 0 & 0 & 0 & 1 \end{pmatrix}. \quad (3.7.8)$$

The first row of C_1 is set to zero because the failed first sensor now outputs only noise.

The objective is to find a torque signal $w(t)$ of minimum energy (optimal proper auxiliary signal v of smallest L^2 norm) so that the failure of the sensor can be detected, and to construct the associated hyperplane test. Scilab programs in chapter 6 are used to construct the solution of this problem over the interval $[0, 36]$. The optimal auxiliary signal is illustrated in figure 3.7.2.

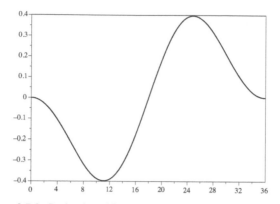

Figure 3.7.2 Optimal auxiliary signal for the suspension problem.

The h vector of the hyperplane test in this case has two components. They are illustrated in figures 3.7.3 and 3.7.4. Note that the components are graphed at very different scales. The hyperplane test greatly emphasizes the output of the first sensor. This is to be expected since the failure is in the first sensor so the first output carries more information. Note, however, that there is a component from the second sensor also.

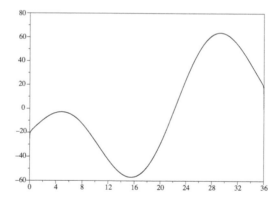

Figure 3.7.3 The first component of the h vector of the hyperplane test.

Simulations were performed in Scicos, which is an object oriented simulation package that comes with Scilab. Scicos has the same functionality that the simulation package SIMULINK has in MATLAB. The noise used in the simulation was a Gaussian white noise. The initial condition was taken as zero. The results of the simulation are illustrated in figure 3.7.5 for when the system is the normal system and in figure 3.7.6 for when it is the failed system. In each case, the first two curves are the outputs of the two sensors and the last curve is the optimal auxiliary signal.

Figure 3.7.7 gives the graph of $\int_0^t h(\tau)^T y(\tau)d\tau$. The threshold for the hyperplane test is $\int_0^T h(\tau)^T y^*(\tau)dt \approx 2$. At the end of the test period ($T = 36$), the normal

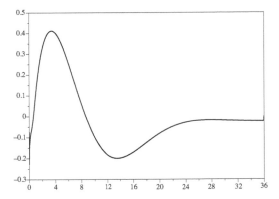

Figure 3.7.4 The second component of the h vector of the hyperplane test.

system yields a value of 4.8, which is greater than the threshold 2, while the failed system yields 0.1. Looking at figure 3.7.7, it is natural to ask whether it is often possible to arrive at these decisions prior to the end of the test period. The answer is yes. Early detection is discussed in section 4.4.

The auxiliary signal is designed to counteract the worst case noise. In many examples it is possible to compute the worst case noise. It quickly becomes apparent that for a given problem the worst case noise has a very definite shape and is not "random appearing." In some sense the worst case noise is often unlikely. This is illustrated by simulation results which show that the brute force application of the failure detection method proposed here using the a posteriori computed energy is very conservative when dealing with white noise. (In this simulation, the variance of the noise is selected such that at the end of each simulation the noise energy over $[0, T]$ is approximately 1.) By "conservative" we mean that the auxiliary signal is larger than would be necessary if we knew that the noise was white with variance less than or equal to 1. Of course, in applications, noise is often not white and "whiteness" is assumed for theoretical reasons.

As we saw in the previous chapter, it is possible to give a statistical interpretation to our approach in such cases. In particular, bounding the L^2 norm of the signal should be interpreted as excluding trajectories that have a likelihood below a certain level. This means that instead of choosing the power of the white noise such that the L^2 norm of the noise over $[0, T]$ falls below 1, we can use the value of the power as a regulation parameter where we adjust the probability of false alarm.

A tenfold increase in the amplitude of the white noise yields results for which it is difficult to detect failure by visual inspection. The hyperplane test, however, gives correct detection in most cases. A typical simulation result with a tenfold increase is given in figures 3.7.8 and 3.7.9. Figure 3.7.8 is for the normal system and figure 3.7.9 for the failed system.

Figure 3.7.10 shows the hyperplane test detecting failure correctly for a number of random simulations with such noise powers. In fact, it made a correct decision for every simulation. However, with this much noise energy, it would be easy to choose

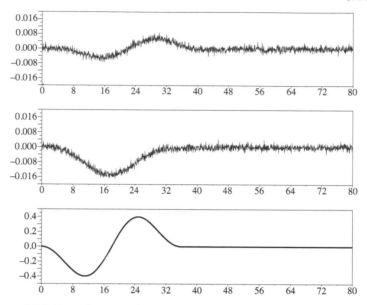

Figure 3.7.5 Normal system: the outputs of the sensors and the auxiliary signal.

a noise trajectory such that the hyperplane test would give the wrong answer. It is the white noise nature of the noise signal here that makes the test work with such high-energy signals (way beyond what is predicted by the theory).

Note that in this example it would be easy for the noise, especially the noise on the dynamics, to not be white. For example, pavement often has sections where there are ripples or a sequence of cracks between sections of pavement, in which case the noise could have periodic components.

3.7.3 Model Uncertainty

Let us now consider again the same suspension problem but with the following modification in (3.7.1b):
$$-Mr\sin(b)\ddot{y} + ((m+M)r^2 + J)\ddot{\theta} = (1+\delta)w, \qquad (3.7.9)$$
where $|\delta| < \bar{\delta}$ is a possibly time-varying model uncertainty. This corresponds to uncertainty in the gain of the torque control input w. Following the techniques presented in the previous chapter (see (2.3.21)), we can construct the two linear models corresponding to this problem. The resulting models are very similar to those of the previous example except for the M matrices, which become
$$M_i = \begin{pmatrix} 0 & 0 \\ \mathcal{P}^{-1}\begin{pmatrix} 0 \\ 1 \end{pmatrix} & 0 \end{pmatrix}, \qquad (3.7.10)$$
$i = 0, 1$, and the presence of G and H matrices
$$G_i = \begin{pmatrix} 0 & 0 & 0 & 0 \end{pmatrix}, \ H_i = \bar{\delta}. \qquad (3.7.11)$$

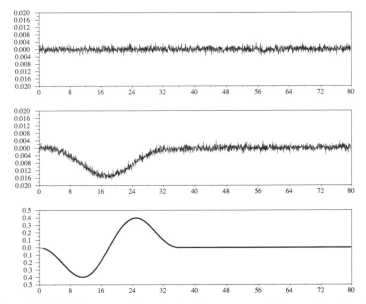

Figure 3.7.6 Failed system: the outputs of the sensors and the auxiliary signal.

The Scilab program in section 6.2 of chapter 6 can be used to construct the optimal auxiliary signal v and the associated hyperplane test signals. For this particular problem, the shape of the resulting optimal auxiliary signal is not very different from the one obtained in the previous case where added uncertainty was considered. However, the separability index has decreased. This is to be expected. Since the model uncertainty is in the B_i matrices and not in the dynamics, a stronger input signal is needed to overcome the possibly lower norm of the input to output function.

Note that in this example, we can remove θ from the state and rewrite the linear system as a three-dimensional system. The reason is that θ does not appear explicitly in any of the equations. This does simplify the problem. The advantage of the current formulation is that it is easy to consider a number of different problems. For example, it is easy to consider the situation where, instead of measuring $\dot{\theta}$, we measure θ. For that, we simply need to modify C_0 and C_1 by exchanging their second and fourth columns.

3.7.4 General Cost Formulation

We can also illustrate the use of more general cost functions of type (3.5.1a) on this example. Suppose that not only do we want to have the test signal small but, if a failure has not occurred, we also want the system to have returned close to its original steady operating point at the end of the test period.

This can be done by using a nominal model of the normal system for (3.5.1b), setting $U = 0$ in (3.5.1a), and using W in (3.5.1a) to put a weight on the final value of the nominal system. By nominal system, we mean the model of the unfailed

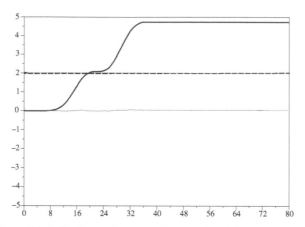

Figure 3.7.7 The hyperplane test allows perfect failure detection.

system with zero initial condition and no noise. This gives the effect of the test signal on the unfailed system if there had been no noise. This amounts to taking $F = A_0$ and $G = B_0$ in (3.5.1b).

The choice of the weight to be placed on the final condition, however, should be made with care. Let us suppose that we simply set $W = cI$ where c is a scalar. Figure 3.7.11 shows the results using the nominal model and different values of c. The top is $c = 0$, then a small c, and finally, at the bottom, large c. The left column shows the graphs of the minimum proper auxiliary signal v and the right shows the response of the nominal model to that v.

It is clear that the final weighting affects primarily one of the components of the state, which is θ. Indeed, since the auxiliary signal starts by speeding up the vehicle before slowing it down, if $c = 0$, then the distance traveled by the car during the test period is more than it would have been if no auxiliary signal were used. When $c > 0$, the extra distance traveled is reduced. This is caused by a larger negative torque in the second half of the test period. The L^2 norm of v increases as c is increased. In addition, in the lowest graph we see that the effort to make $\theta(T)$ small has actually made the final values of the other variables worse then they were with $c = 0$.

But the actual distance covered has no importance in this problem. What matters is the vertical position of the suspension, its speed, and the speed of the vehicle. So not only is the $W = cI$ weight useless as far as the final value of θ is concerned, it makes things worse.

The correct way to choose W would be to put a weight on y and $\dot{y}$. It would be reasonable also to put a weight on $\dot{\theta}$ so that the test does not affect the speed of the vehicle too much. Thus W should be in the form

$$W = \begin{pmatrix} w_1 & 0 & 0 & 0 \\ 0 & w_2 & 0 & 0 \\ 0 & 0 & 0 & 0 \\ 0 & 0 & 0 & w_4 \end{pmatrix}, \qquad (3.7.12)$$

where $\{w_1, w_2, w_4\}$ are positive numbers chosen based on design considerations.

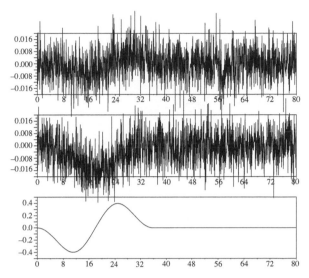

Figure 3.7.8 Normal system: the outputs of the sensors and the auxiliary signal.

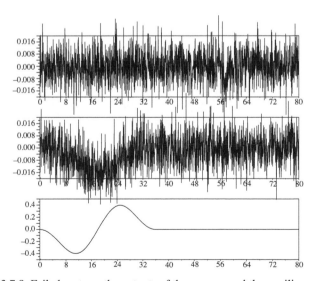

Figure 3.7.9 Failed system: the outputs of the sensors and the auxiliary signal.

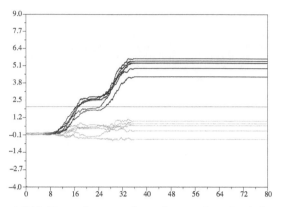

Figure 3.7.10 Hyperplane tests for various random simulation runs.

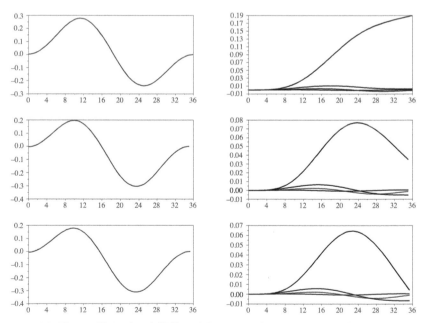

Figure 3.7.11 The auxiliary signal (left) and the state trajectory of the nominal model (right)
                         for three values of c.

3.8 ASYMPTOTIC BEHAVIOR

In our approach to auxiliary signal design, we are interested in constructing optimal auxiliary signals over short periods of time. It is, however, instructive and to a certain degree useful to consider what happens when T is large by studying the asymptotic behavior as T goes to infinity. This problem was considered in [66]. We are not going to redo the theory here, but simply recall some of the conclusions and show how they can be used in our setting.

By running random examples, we can observe the following. There is a unique frequency associated with each problem, which we call the dominant frequency. For large T, the dominant frequency constitutes the main component of the optimal proper auxiliary signal. By "large T" we mean $1/T$ at least a few times less than the dominant frequency. The dominant frequency does not depend on T.

By extending the results of [66] to the continuous-time case, we can show that the dominant frequency can be obtained from the Hamiltonian matrix associated with the system (3.3.34):

$$\Omega(\lambda, \beta) = \begin{pmatrix} \Omega_{11} & \Omega_{12} \\ \Omega_{21} & \Omega_{22} \end{pmatrix}. \tag{3.8.1}$$

In particular, as T goes to infinity, λ_β converges to λ_β^∞ where λ_β^∞ is the smallest value of λ for which $\Omega(\lambda, \beta)$ has no eigenvalue on the imaginary axis (assuming it does not have any eigenvalue on the imaginary axis for $\lambda = \infty$). This computation is very much reminiscent of γ iteration in H_∞ control. By maximizing λ_β^∞ over β, the limiting values of λ^* and β^*, denoted, respectively, λ_∞^* and β_∞^*, can be obtained. The Hamiltonian matrix $\Omega(\lambda_\infty^*, \beta_\infty^*)$ has an eigenvalue on the imaginary axis. If $j\omega_\infty^*$ is one such eigenvalue, then ω_∞^* is the dominant frequency.

For large T, λ_β^∞ is a good approximation of λ_β. It actually overbounds it because the separability index is monotonically nondecreasing in T (see [66] for more details on that). Thus β_∞^* is a good approximation for β^*.

In the scalar case, the optimal auxiliary signal can be approximated by

$$v_{\infty,T}^* = \alpha \sin\left(\pi \frac{t}{T}\right) \cos(\omega_\infty^* t + \phi), \tag{3.8.2}$$

where α is chosen such that $\|v_{\infty,T}^*\| = 1/\gamma_\infty^*$ and γ_∞^* is the limiting separability index given by

$$\gamma_\infty^* = \sqrt{\lambda_\infty^*}. \tag{3.8.3}$$

Here ϕ is an arbitrary phase value.

Figure 3.8.1 shows the separability index as a function of T for a randomly generated example. Note that by $T = 18$ it appears close to its limiting value.

Figure 3.8.2 shows the graphs of λ_β for $T = 18$ and λ_β^∞. Figure 3.8.3 shows the optimal solution v on the infinite interval and the one on $[0, T]$. In this case, since $1/T$ is a few times less than the dominant frequency, we see that $v_{\infty,T}^*$ is a good approximation to v and the performance of $v_{\infty,T}^*$ is satisfactory.

The computation of $v_{\infty,T}^*$ requires much less computational effort than the computation of the true optimal proper auxiliary signal. That is not such an important

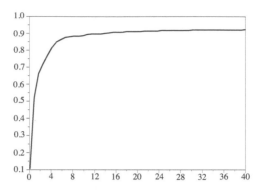

Figure 3.8.1 Separability index as a function of T.

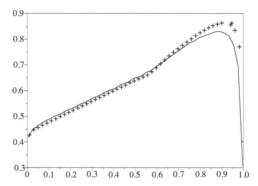

Figure 3.8.2 λ_β for $T = 18$ (solid line) and λ_β^∞ (dotted line).

factor because this computation is done off-line. But $v_{\infty,T}^*$ also needs almost no storage compared to the optimal proper auxiliary signal, which needs to be coded using, for example, splines. That is why the asymptotic solution can be of use in some situations. Even when $v_{\infty,T}^*$ cannot be used, knowing the dominant frequency can be useful for choosing the right coding for the true optimal proper auxiliary signal.

3.9 USEFUL RESULTS

3.9.1 Optimization of the Maximum of Quadratic Functionals

In our development, the inf-max of the noise measures plays a fundamental role. The actual form of the noise measures varies but the following situation covers all cases. We consider the following optimization problem:

$$c = \inf_x \max(G_0(x), G_1(x)), \tag{3.9.1}$$

where the $G_i(x)$ are quadratic functionals and $x \in \mathcal{X}$ where $\mathcal{X}$ is a real Hilbert space. That is, $\mathcal{X}$ has an inner product $\langle,\rangle$ and contains all of its limit points. In

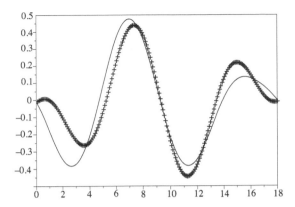

Figure 3.8.3   Optimal proper auxiliary signal for $T = 18$ (solid line) and $v_{\infty,T}^*$ (dotted line).

general, $\mathcal{X}$ will be a direct sum of an L^2 space (which constrains the noise ν) and a finite-dimensional space (which contains $x(0)$). The functionals may be written

$$G_i(x) = \langle P_i x, x \rangle + \langle x, q_i \rangle + r_i, \qquad (3.9.2)$$

where P_i is a bounded symmetric operator, $q_i \in \mathcal{X}$, and r_i is a real number. However, sometimes $\mathcal{X}$ will be finite dimensional and then (3.9.2) becomes

$$G_i(x) = x^T P_i x + q_i^T x + r_i, \qquad (3.9.3)$$

$i = 0, 1$, for matrices P_i, vectors q_i, and scalars r_i.

We will need the following facts.

Lemma 3.9.1 *For any quadratic functional $G(x)$, if $G(x_0) = \alpha_0 > \alpha_1 = G(x_1)$, then there is a closed interval $[t_0, t_1]$ and a continuous function $x(t)$ on this interval such that $G(x(t_i)) = \alpha_i$ for $i = 0, 1$, $x(t_0) = x_0$, and $G(x(t))$ is monotonically nonincreasing on the interval.*

Proof. Note that we do not assert that $x(t_1) = x_1$. In the actual use of this result later we really only need to be able to take $G(x(t_i))$ arbitrarily close to α_i. Thus we can assume without loss of generality that 0 is an isolated point of the spectrum of P. If G is nonnegative and $G(x_0) > 0$, then the proof follows from just having $x(t)$ rescale x_0. For the general case, note that P_i being symmetric means that we may decompose $\mathcal{X}$ into an orthogonal sum $\mathcal{X}_1 \oplus \mathcal{X}_2 \oplus \mathcal{X}_3$, and relative to this direct sum we have $P_i = P_+ \oplus -P_- \oplus 0$ where P_+, P_- are positive definite, and $x = x_+ \oplus x_- \oplus x_0$, $q = q_+ \oplus q_- \oplus q_0$. By the change of variables $x = z + a$ we may assume $q_+ = 0, q_- = 0$. Thus (3.9.2) becomes $G(x) = \langle P_+ x_+, x_+ \rangle - \langle P_- x_-, x_- \rangle + \langle x_0, q_0 \rangle + r$. The lemma now follows by considering the individual summands and scaling them separately. $\qquad\square$

Lemma 3.9.2 *Suppose c is defined by (3.9.1). Let $\mathcal{S} = \{x | G_0(x) = G_1(x)\}$. If $c \geq 0$, then at least one of the following statements holds:*

1. *c is the global minimum of $G_0(x)$,*

$$c = \min_x G_0(x); \tag{3.9.4}$$

2. *c is the global minimum of $G_1(x)$,*

$$c = \min_x G_1(x); \tag{3.9.5}$$

3. *c is the minimum on S,*

$$c = \inf_{x \in S} G_0(x) = \inf_{x \in S} G_1(x). \tag{3.9.6}$$

Note that the infimum in (3.9.6) cannot always be replaced with a minimum even if $c \geq 0$, as can be seen from the example $c = \inf_{x_1, x_2} \max(x_1^2, 1 - x_1 x_2)$. Clearly here $c = 0$ but zero is not achieved by any (x_1, x_2).

Proof. Suppose that $c \geq 0$. Let δ be the infimum in (3.9.6). Clearly $c \leq \delta$. If $c = \delta$ we are done since then (3.9.6) holds. Suppose then that $c < \delta$. Let $\{x_j\}$ be a sequence such that $\max\{G_0(x_j), G_1(x_j)\}$ decreases monotonically to c. Let

$$S_1 = \{x : G_0(x) > G_1(x)\}, \quad S_2 = \{x : G_1(x) > G_0(x)\}.$$

Since $c < \delta$ there must be a subsequence of $\{x_j\}$ which lies in S_1 or S_2. We may assume that it is in S_1. Thus we have $G_0(x_j) \to c$. If c is the global minimum of G_0, then we are done. Suppose this is not the case. Then there is an $\hat{x} \in S_1$ with $c \leq G_0(\hat{x}) < \delta$ and an $\tilde{x} \in S_2$ with $G_0(\tilde{x}) < c \leq G_0(\hat{x})$. Applying lemma 3.9.1, we get an $x(t)$ on $[0, 1]$ such that $x(0) = \hat{x}$ and $G_0(x(1)) = G_0(\tilde{x})$. Clearly $x(1) \in S_2$. But now consider the continuous scalar function $b(t) = G_0(x(t)) - G_1(x(t))$. We have $b(0) > 0$ and $b(1) < 0$ so there is a value $\hat{t}$ so that $b(\hat{t}) = 0$. But then this $x(\hat{t})$ is in S and $G_0(x(\hat{t})) < \delta$ because $G_0(x(t))$ decreases. But this contradicts the definition of δ. $\qquad\square$

Lemma 3.9.3 *If $c \geq 0$, then there exists a $\beta \in [0, 1]$ such that*

$$\beta G_0(x) + (1 - \beta)G_1(x) \geq 0 \tag{3.9.7}$$

for all x.

Proof. We can suppose that there exists an x such that $G_1(x) < 0$ (otherwise, $G_1(x) \geq 0$ for all x and we can take $\beta = 0$). Also, clearly, for any x, if $G_1(x) < 0$, then $G_0(x) \geq 0$ since otherwise $c < 0$.

We can now use theorem 4.2.1 of [73] which states that for two quadratic functionals $\mathcal{G}_0(x)$ and $\mathcal{G}_1(x)$, if there exists $\bar{x}$ such that $\mathcal{G}_1(\bar{x}) > 0$, then the following conditions are equivalent:

1. $\mathcal{G}_0(x) \geq 0$ for all x such that $\mathcal{G}_1(x) \geq 0$;

2. There exists a constant $\tau \geq 0$ such that for all x we have

$$\mathcal{G}_0(x) - \tau \mathcal{G}_1(x) \geq 0. \tag{3.9.8}$$

Now let $\mathcal{G}_1(x) = -G_1(x)$ and $\mathcal{G}_0(x) = G_0(x)$. Then from the above result it follows that there exists a constant $\tau \geq 0$ such that

$$G_0(x) + \tau G_1(x) \geq 0 \tag{3.9.9}$$

for all x. If we let

$$\tau = \frac{1 - \beta}{\beta} \tag{3.9.10}$$

we obtain (3.9.7), which is the desired result. $\square$

Theorem 3.9.1 *If $c \geq 0$, then*

$$c = \max_{\beta \in [0,1]} \inf_x \beta G_0(x) + (1 - \beta)G_1(x). \tag{3.9.11}$$

Proof. Since $\max \inf \leq \inf \max$, we know that

$$c \geq \max_{\beta \in [0,1]} \inf_x \beta G_0(x) + (1 - \beta)G_1(x). \tag{3.9.12}$$

Thus if condition (3.9.4) holds, then the equality is obtained by setting $\beta = 1$. Similarly, if (3.9.5) holds, then the equality is obtained by setting $\beta = 0$. So the only case that we need to consider is when

$$c = \inf_{x \in S} G_0(x) = \inf_{x \in S} G_1(x). \tag{3.9.13}$$

Let $\mathcal{B}$ denote the set of all β for which $\beta G_0(x) + (1 - \beta)G_1(x)$ is convex in x. This set is not empty because it contains at least one element, as shown in lemma 3.9.3, and it is closed. Thanks to (3.9.13), we have

$$\begin{aligned}
c &= \inf_x \max_{\beta \in [0,1]} \beta G_0(x) + (1 - \beta)G_1(x) \\
&= \inf_{x \in S} \max_{\beta \in [0,1]} \beta G_0(x) + (1 - \beta)G_1(x) \\
&= \inf_x \max_{\beta \in \mathcal{B}} \beta G_0(x) + (1 - \beta)G_1(x) \\
&= \max_{\beta \in \mathcal{B}} \inf_x \beta G_0(x) + (1 - \beta)G_1(x). \tag{3.9.14}
\end{aligned}$$

But $\inf_x \beta G_0(x) + (1 - \beta)G_1(x)$ is $-\infty$ if β is not in $\mathcal{B}$, proving (3.9.11). $\square$

Corollary 3.9.1 *If $c > -\infty$, then (3.9.11) holds.*

Proof. Suppose $c = -p < 0$. Then simply let $G_i(x) = p + G_i(x), i = 0, 1$. Then apply theorem 3.9.1. $\square$

Thus we have shown that the order of the infimum and maximum can be exchanged in problem (3.9.1), when c is finite. We have also shown that maximum of the infimum is $-\infty$ when c is $-\infty$. This shows that we can exchange the order of the maximum and the infimum in every case.

Lemma 3.9.4 *Let $\mathcal{B}$ be the set of β for which $\beta G_0(x) + (1 - \beta)G_1(x)$ is convex in x. Then $\mathcal{B}$ is a closed interval inside $[0, 1]$.*

Proof. It is easy to see that $\beta \in \mathcal{B}$ if and only if

$$M(\beta) \triangleq \beta P_0 + (1 - \beta)P_1 \geq 0. \tag{3.9.15}$$

Clearly then $\mathcal{B}$ is closed. Now let $0 \leq \lambda \leq 1$, and let β_1 and β_2 be in $\mathcal{B}$. We can see that $\lambda\beta_1 + (1 - \lambda)\beta_2$ is in $\mathcal{B}$ by noting that

$$M(\lambda\beta_1 + (1 - \lambda)\beta_2) = \lambda M(\beta_1) + (1 - \lambda)M(\beta_2) \geq 0. \tag{3.9.16}$$

Thus $\mathcal{B}$ is convex. $\qquad\qquad\square$

The solution to the optimization problem (3.9.1) can thus be expressed as

$$c = \max_{\beta \in \mathcal{B}} \inf_x \beta G_0(x) + (1 - \beta)G_1(x). \tag{3.9.17}$$

The minimization problem is a standard LQ problem, and the maximization is that of a concave function over a finite interval.

Note also that $\mathcal{B}$ being not empty does not imply that c is not $-\infty$. For that we need an additional assumption of nonsingularity.

Definition 3.9.1 *The optimization problem (3.9.1) is called nonsingular if there exists a $\beta \in [0, 1]$ such that $\beta G_0(x) + (1 - \beta)G_1(x)$ is strictly convex.*

Clearly, in the finite-dimensional case, nonsingularity is equivalent to the existence of $\beta \in [0, 1]$ such that

$$\beta P_0 + (1 - \beta)P_1 > 0. \tag{3.9.18}$$

Lemma 3.9.5 *If the optimization problem (3.9.1) is nonsingular, then $c > -\infty$.*

Proof. Let β be such that (3.9.18) is satisfied. Then the optimization problem

$$\inf_x \beta G_0(x) + (1 - \beta)G_1(x) \tag{3.9.19}$$

has a finite solution because we have a strictly convex quadratic functional. Thus $c > -\infty$. $\qquad\qquad\square$

Note that nonsingularity is not a necessary condition for $c > -\infty$. It is a sufficient condition. However, it is a necessary and sufficient condition for $c > -\infty$ to hold for all q_i and r_i. Also note that in the nonsingular case $\mathcal{B}$ is not reduced to a single point. Singular problems are nongeneric. That is, if one were to randomly generate an example using a probablilty distribution without point masses (such as a normal or Gaussian distribution) the probability of a singular problem would be zero. Such comments about genericness always have the two caveats that many problems have a structure to them so the entries are not independent and if a nonsingular problem is close to being singular we may still have practical difficulties.

3.9.2 Construction of the Auxiliary Signal

Consider the optimization problem

$$d = \max_{v \neq 0} \frac{\inf_x \max(G_0(x, v), G_1(x, v))}{|v|_Q^2}, \tag{3.9.20}$$

where $G_i(x, v)$ are quadratic functionals in $(x, v) \in \mathcal{X} \times \mathcal{V}$, $\mathcal{X}$ and $\mathcal{V}$ are real linear vector spaces, and $|\cdot|_Q$ is a norm defined on $\mathcal{V}$. In the case where these spaces are finite dimensional, we have for some matrices P_i, Q_i, and R_i,

$$G_i(x) = x^T P_i x + 2v^T Q_i^T x + v^T R_i v \qquad (3.9.21)$$

for $i = 0, 1$.

This problem is of course closely related to the optimization problem considered in the previous section; for example, in the finite-dimensional case this can be seen by letting

$$2Q_i v = q_i, \quad v^T R_i v = r_i. \qquad (3.9.22)$$

Lemma 3.9.6 *Let*

$$J(v) = \inf_x \max(G_0(x, v), G_1(x, v)). \qquad (3.9.23)$$

Then $J(v)$, if it is not $-\infty$, is quadratic in v.

Proof. We have to show that for all scalar constants a,

$$J(av) = a^2 J(v). \qquad (3.9.24)$$

Suppose $a \neq 0$; then by letting $y = x/a$, we obtain

$$J(av) = \min_y \max(G_0(ay, av), G_1(ay, av))$$

$$= \min_y \max(a^2 G_0(y, v), a^2 G_1(y, v)) = a^2 J(v). \qquad (3.9.25)$$

Clearly, since $J(v)$ is not $-\infty$ for some v, there exists a β such that $\beta G_0(x, v) + (1 - \beta)G_1(x, v)$ is convex. Thus $J(0) = 0$. $\qquad \square$

Note that lemma 3.9.6 does not imply that $J(v)$ is a quadratic functional. That is because $J(v_1)$ may exist but not $J(v_2)$. For that, we need an additional assumption, nonsingularity. We shall suppose from here on that our problem is nonsingular.

Because of nonsingularity,

$$J(v) = \mathcal{S}(v, v) \qquad (3.9.26)$$

for some bilinear form on $\mathcal{V} \times \mathcal{V}$. So d can be expressed as follows,

$$d = \max_{v \neq 0} \frac{\mathcal{S}(v, v)}{|v|_Q^2}. \qquad (3.9.27)$$

In the finite-dimensional case, there exists a symmetric matrix S so that $\mathcal{S}(v, v) = v^T S v$, i.e.,

$$J(v) = v^T S v, \qquad (3.9.28)$$

and d can be expressed as

$$d = \max_{v \neq 0} \frac{v^T S v}{v^T Q v}. \qquad (3.9.29)$$

This value of d, which is the largest eigenvalue of SQ^{-1}, can be obtained as the smallest δ for which the optimization problem

$$e = \max_v v^T S v - \delta v^T Q v \qquad (3.9.30)$$

has zero solution. The optimizing v is then given as a nontrivial solution of

$$Sv = dQv. \tag{3.9.31}$$

Using the results of the previous section, we can express this optimization problem as follows:

$$e = \max_v \max_{\beta \in \mathcal{B}} \inf_x \beta G_0(x, v) + (1 - \beta)G_1(x, v) - \delta v^T Qv. \tag{3.9.32}$$

Let us define

$$L_{\beta, \delta}(x, v) = \beta G_0(x, v) + (1 - \beta)G_1(x, v) - \delta v^T Qv. \tag{3.9.33}$$

Clearly $L_\beta(x, v)$ is quadratic in (x, v).

From (3.9.32) and (3.9.33), exchanging the order of the two maximizations, we obtain

$$e = \max_{\beta \in \mathcal{B}} \max_v \inf_x L_{\beta, \delta}(x, v). \tag{3.9.34}$$

Theorem 3.9.2 *Let*

$$M(\beta, \delta) = \begin{pmatrix} \beta P_0 + (1 - \beta)P_1 & \beta Q_0^T + (1 - \beta)Q_1^T \\ \beta Q_0 + (1 - \beta)Q_1 & \beta R_0 + (1 - \beta)R_1 - \delta Q \end{pmatrix} \tag{3.9.35}$$

and let d_β, for $\beta \in \mathcal{B}$, be the largest δ such that

$$\mathrm{Ker}(M(\beta, \delta)) \not\subset \mathrm{Im}\begin{pmatrix} I \\ 0 \end{pmatrix}. \tag{3.9.36}$$

Then

$$d = \max_{\beta \in \mathcal{B}} d_\beta. \tag{3.9.37}$$

This result is obtained directly from the optimality principle.

Note that if β is such that $\beta P_0 + (1 - \beta)P_1 > 0$, which is the case for all β in the interior of $\mathcal{B}$ (if not empty, i.e., if the problem is nonsingular), then the condition (3.9.36) is equivalent to invertibility of $M(\beta, \delta)$. Since d_β is continuous in β, this gives us a straightforward optimization algorithm.

Corollary 3.9.2 *Let d be defined as in (3.9.37), and let β^* be the maximizing β. Then any optimizing v in (3.9.29) satisfies*

$$M(\beta^*, d)\begin{pmatrix} x \\ v \end{pmatrix} = 0. \tag{3.9.38}$$

3.9.3 Continuous-time Problem and the Riccati Equation

Lemma 3.9.7 *Suppose the Hamiltonian two-point boundary value system*

$$\dot{\xi} = \begin{pmatrix} A & Q \\ R & -A^T \end{pmatrix}\xi, \tag{3.9.39a}$$

$$0 = V_0\xi(0) + V_T\xi(T), \tag{3.9.39b}$$

where

$$V_0 = \begin{pmatrix} I & -P_0 \\ 0 & 0 \end{pmatrix}, \quad V_T = \begin{pmatrix} 0 & 0 \\ 0 & I \end{pmatrix}, \tag{3.9.40}$$

is well posed (i.e., has a unique zero solution) for all $T \in [0, \bar{T})$ but is not well posed for $T = \bar{T}$. Then the solution of the Riccati equation

$$\dot{P} = AP + PA^T - PRP + Q, \quad P(0) = P_0, \tag{3.9.41}$$

diverges at $t = \bar{T}$.

Proof. Let Ψ be the system matrix, i.e., $\Psi(0) = I$, and

$$\frac{d}{dt} \begin{pmatrix} \Psi_1 & \Psi_2 \\ \Psi_3 & \Psi_4 \end{pmatrix} = \begin{pmatrix} A & Q \\ R & -A^T \end{pmatrix} \begin{pmatrix} \Psi_1 & \Psi_2 \\ \Psi_3 & \Psi_4 \end{pmatrix}. \tag{3.9.42}$$

The well-posedness of the two-point boundary value system over $[0, T]$ can then be characterized in terms of the invertibility of the following matrix:

$$V_0 + V_T \Psi(T) = \begin{pmatrix} I & -P_0 \\ \Psi_3(T) & \Psi_4(T) \end{pmatrix}. \tag{3.9.43}$$

After straightforward manipulation it can be shown that invertibility of (3.9.43) is equivalent to invertibility of $\Psi_3(T)P_0 + \Psi_4(T)$.

Let

$$Y(t) = \begin{pmatrix} Z(t) \\ M(t) \end{pmatrix} = \Psi(t) \begin{pmatrix} P_0 \\ I \end{pmatrix} = \begin{pmatrix} \Psi_1(t)P_0 + \Psi_2(t) \\ \Psi_3(t)P_0 + \Psi_4(t) \end{pmatrix}. \tag{3.9.44}$$

Then we know that $M(t)$ is invertible for $t < \bar{T}$ and singular for $t = \bar{T}$. It is a standard but somewhat long and delicate argument to verify that

$$P(t) = Z(t)M(t)^{-1} \tag{3.9.45}$$

satisfies the Riccati equation (3.9.41). To see that $P(t)$ also diverges at $t = \bar{T}$ note that there is a vector a, $a \neq 0$, such that $M(\bar{T})a = 0$ since $M(\bar{T})$ is singular. From the basic theory of linear differential equations we have $Y(t)a \neq 0$ for all t since it is nonzero at $t = 0$. Since $M(T)a = 0$ it follows that $Z(T)a \neq 0$. Also $Z(t)a$ is continuous at $\bar{T}$ since Ψ is. But $Z(T)a = Z(t)M(t)^{-1}[M(t)a]$ and $\lim_{t \to \bar{T}-} M(t)a = 0$ gives that $Z(t)M(t)^{-1}$ has a singularity at $t = \bar{T}$ since $Z(t)M(t)^{-1}$ is unbounded as t approaches $\bar{T}$. $\square$

Lemma 3.9.7 is sufficient for our purposes. More general results and discussion can be found, for example, in [74].

Example 3.9.1 *Consider the case where*

$$A = 0, \quad Q = -\alpha, \quad R = 1. \tag{3.9.46}$$

Assume first that $\alpha > 0$. To test the well-posedness of (3.9.39), we need to evaluate the matrix in (3.9.43). Note that

$$\Psi(T) = \exp\left(\begin{pmatrix} 0 & -\alpha \\ 1 & 0 \end{pmatrix} T \right)$$

$$= I + \begin{pmatrix} 0 & -\alpha \\ 1 & 0 \end{pmatrix} T + \frac{1}{2} \begin{pmatrix} 0 & -\alpha \\ 1 & 0 \end{pmatrix}^2 T^2 + \frac{1}{6} \begin{pmatrix} 0 & -\alpha \\ 1 & 0 \end{pmatrix}^3 T^3 + \cdots$$

$$= \begin{pmatrix} \cos(\sqrt{\alpha}T) & -\sqrt{\alpha}\sin(\sqrt{\alpha}T) \\ \frac{1}{\sqrt{\alpha}}\sin(\sqrt{\alpha}T) & \cos(\sqrt{\alpha}T) \end{pmatrix}. \tag{3.9.47}$$

Then

$$V_0 + V_T \Psi(T) = \begin{pmatrix} 1 & -p_0 \\ \dfrac{1}{\sqrt{\alpha}} \sin(\sqrt{\alpha}T) & \cos(\sqrt{\alpha}T) \end{pmatrix}, \tag{3.9.48}$$

which clearly is noninvertible if and only if

$$\cos(\sqrt{\alpha}T) + \frac{p_0}{\sqrt{\alpha}} \sin(\sqrt{\alpha}T) = 0. \tag{3.9.49}$$

On the other hand, the Riccati equation (3.9.41) becomes

$$\dot{p} = -p^2 - \alpha. \tag{3.9.50}$$

It is easy to show by looking at the phase portrait that if $\alpha > 0$ and $p(0) > 0$, then p becomes zero in finite time and then goes to $-\infty$ at a later finite time. This illustrates the point made earlier about P being singular prior to having a singularity.

Equation (3.9.50) can be expressed as

$$\frac{dp}{p^2 + \alpha} = -dt. \tag{3.9.51}$$

By integrating both sides and after some straightforward algebra using $p(0) = p_0$, we get

$$p = \sqrt{\alpha} \tan\left(-\sqrt{\alpha}t + \tan^{-1}\left(\frac{p_0}{\sqrt{\alpha}}\right)\right). \tag{3.9.52}$$

Thus the Riccati equation diverges at $t = T$ if T satisfies the relation

$$\begin{aligned} 0 &= \cos\left(-\sqrt{\alpha}T + \tan^{-1}\left(\frac{p_0}{\sqrt{\alpha}}\right)\right) \\ &= \frac{1}{\sqrt{\dfrac{p_0^2}{\alpha} + 1}} \left(\cos\left(\sqrt{\alpha}T + \frac{p_0}{\sqrt{\alpha}} \sin(\sqrt{\alpha}T)\right)\right). \end{aligned} \tag{3.9.53}$$

Clearly (3.9.49) and (3.9.53) are consistent as predicted by lemma 3.9.7.

Now suppose that $\alpha = 0$. Then

$$\Psi(T) = \exp\left(\begin{pmatrix} 0 & 0 \\ 1 & 0 \end{pmatrix} T\right) = \begin{pmatrix} 1 & 0 \\ T & 1 \end{pmatrix}. \tag{3.9.54}$$

Thus

$$V_0 + V_T \Psi(T) = \begin{pmatrix} 1 & -p_0 \\ T & 1 \end{pmatrix}, \tag{3.9.55}$$

which is invertible for all T since its determinant is $1 + p_0 T$.

On the other hand, the Riccati equation (3.9.41) becomes

$$\dot{p} = -p^2, \tag{3.9.56}$$

whose solution is either $p = 0$ or

$$p = \frac{1}{t + p_0^{-1}}, \tag{3.9.57}$$

which exists and is nonzero for all $t > 0$. *Thus (3.9.49) and (3.9.53) are again consistent, as predicted by lemma 3.9.7.*

If $\alpha < 0$, *then the Riccati equation is*

$$\dot{p} = -p^2 + b^2 \tag{3.9.58}$$

with $b > 0$. *There are two equilibrium (constant) solutions of* $p = b$ *and* $p = -b$. *If* $p(0) > b$, *then* $\lim_{t \to \infty} p(t) = b$ *monotonically, there are no singularities, and* p *is never zero. If* $b > p(0) > -b$, *then again* $\lim_{t \to \infty} p(t) = b$ *so* p *does not have a singularity. It will be singular for some* t *if* $0 \geq p(0) > -b$. *If* $p(0) < -b$, *then* p *has a singularity in finite time but is never singular. This illustrates the important role played by our assumption that* $P(0) > 0$.

In conclusion, we note that if $\alpha < 0$ *and* $p(0) > 0$, *then the solution of the Riccati equation is given by*

$$p = \sqrt{-\alpha} \frac{p_0 \cosh(-\sqrt{-\alpha}t/2) + \sqrt{-\alpha} \sinh(-\sqrt{-\alpha}t/2)}{p_0 \sinh(-\sqrt{-\alpha}t/2) + \sqrt{-\alpha} \cosh(-\sqrt{-\alpha}t/2)} \tag{3.9.59}$$

which does not diverge and the two-point boundary value system (3.9.39) is well posed for all $T > 0$.

Chapter Four

Direct Optimization Formulations

4.1 INTRODUCTION

In chapter 3 we considered the problem of robust multimodel identification in the case when there were two models. One model was for the original, normal system and the other model for the system with a failure. In that chapter an elegant theory and numerical algorithms were presented. In this chapter we shall set up an alternative formulation directly in terms of an optimization problem. While less elegant than the previous development, the new approach allows us to consider a number of more general problems. In particular, we are able to design auxiliary signals and detection tests for problems with more than one type of failure so that there are more than two models, with additional known signals which we call reference signals, with some nonlinearities, and with additional design constraints on the auxiliary signal. Also, in the previous chapters we always had a weight on the uncertainty of the initial state. In this chapter we relax that assumption.

In order to simplify the discussion, we primarily focus on the case when there is additive uncertainty. We first develop our optimization formulation for the two-model case. After some brief comments on the software used for the examples, we then consider the general m-model case and address a number of new problem variations that can be solved using the optimization formulation. We also show how early detection is possible using the hyperplane tests of this chapter and chapter 3. A similar discussion is possible for discrete systems but we do not consider them here.

Some applications are described by infinite-dimensional models such as those involving delays or partial differential equations. The approach of this book can be applied provided finite-dimensional approximations are used. However, there are certain technical issues that need to be addressed when doing these approximations. In section 4.6 of this chapter we will illustrate some of these concerns by considering models with delays. We then consider in section 4.7 the question of how to set the error bounds when working with a real physical system. The approach of section 4.7 can be used with the techniques of either chapter 3 or chapter 4.

In the final section we discuss using the optimization approach for problems with model uncertainty. The optimization approach can be used but there are certain additional computational issues that must be considered.

4.2 OPTIMIZATION FORMULATION FOR TWO MODELS

In chapter 3 we considered the problem of choosing between two models with additive uncertainty. The true system model was supposed to be one of the following two linear models defined on $[0, T]$:

$$\dot{x}_i = A_i x_i + B_i v + M_i \nu_i, \qquad (4.2.1a)$$

$$y = C_i x_i + N_i \nu_i \qquad (4.2.1b)$$

for $i = 0$ and 1. As in chapter 3 it is assumed that v, ν_i are in $L^2[0, T] = L^2$. Then x_i, y will also be in L^2. We assume the N_i have full row rank. If controls are present in the form of feedback loops, then they are included in the A_i in (4.2.1a). Other controls such as reference controls will be discussed later in section 4.5. On-line measured inputs may be dealt with as in section 3.4. The system matrices may depend on t.

As before, for $i = 0, 1$, we consider a perturbation or noise bound of the form

$$S_i(x_i(0), \nu_i) = (x_i(0) - x_{i0})^T \hat{P}_i (x_i(0) - x_{i0}) + \|\nu_i\|^2$$

$$= (x_i(0) - x_{i0})^T \hat{P}_i (x_i(0) - x_{i0}) + \int_0^T |\nu_i(t)|^2 dt < 1. \quad (4.2.2)$$

However, unlike in the previous sections, we also allow $\hat{P}_i = 0$, in which case we sometimes write $S_i(\nu_i)$. In the earlier chapters we used $P_i(0)^{-1}$ for the weight instead of $\hat{P}_i$ because that simplified the remaining formulas. However, in this chapter the notation $\hat{P}_i$ will turn out to result in simpler formulas. In this chapter we also assume that $x_{i0} = 0$ as was done in chapter 3. That is, there is no a priori knowledge about x_i.

Depending on the application there are different ways to measure the auxiliary signal v. Following section 3.5 we use

$$\delta^2(v) = \psi(T)^T W \psi(T) + \int_0^T |v|^2 + \psi^T U \psi \, dt, \qquad (4.2.3a)$$

$$\dot{\psi} = F_0 \psi + G_0 v, \quad \psi(0) = 0, \qquad (4.2.3b)$$

where U, W are positive semidefinite matrices. δ is a norm on L^2. In the important special case when $U = 0$ and $W = 0$, $\delta(v)$ is just the L^2 norm of v and we write $\delta(v) = \|v\|$.

4.2.1 When $\hat{P}_i = 0$

The assumption $\hat{P}_i > 0$ was used in the Riccati equations of chapter 3. However, it was not used in the application of the results of section 3.9. Thus the basic theory from chapter 3 such as exchanging the maximum and minimum and $\phi_\beta(v) \geq 1$ characterizing the properness of the auxiliary signal v, still holds. However, $\hat{P}_i = 0$ does have an impact on the geometry, which in turn can affect the numerical solutions. Before developing the optimization formulations we will comment on this difference.

Assuming that $\hat{P}_i = 0$ means that we are assuming that there is no information prior to $t = 0$, and in particular there is no information on $x_i(0)$. This is a natural assumption to make for some applications. The output sets $\mathcal{A}_i(v)$ and what it means for a signal to be proper are as before. In the earlier chapters we had that $\mathcal{A}_i(v)$ was a bounded, strictly convex set in L^2 for a given L^2 auxiliary signal v when $J_i = I$. That is, (4.2.2) holds. This is no longer the case if $\hat{P}_i = 0$. Suppose that $\hat{P}_i = 0$. Let

$$\mathcal{L}_i(f) = \int_0^t e^{A_i(t-s)} f(s) ds \qquad (4.2.4)$$

be the solution of $\dot{z} = A_i z + f$, $z(0) = 0$. (If A_i is time varying, then $e^{A_i t}$ in (4.2.4) is replaced by the fundamental solution matrix [47] $\Theta(t)$, which is also sometimes called the state transition matrix.) Using (4.2.4), we have that all solutions of (4.2.1) are given by

$$x_i = \mathcal{L}_i(B_i v) + \mathcal{L}_i(M_i \nu_i) + e^{A_i t} \xi_i, \qquad (4.2.5)$$
$$y = C_i \mathcal{L}_i(B_i v) + C_i \mathcal{L}_i(M_i \nu_i) + C_i e^{A_i t} \xi_i + N_i \nu_i, \qquad (4.2.6)$$

where ξ_i is the unknown initial condition for x_i. Thus the output set for each model is the algebraic sum of three sets:

1. $C_i \mathcal{L}_i B_i v$, which is a vector depending linearly on v;

2. $\{(C_i \mathcal{L}_i M_i + N_i)\nu_i : \mathcal{S}_i(\nu_i) < 1\}$, which is an open bounded convex set that is symmetric about the origin;

3. $\{C_i e^{A_i t} \xi_i : \xi_i \in R^n (\text{ or } C^n)\}$, which is a finite-dimensional subspace of L^2.

The output sets $\mathcal{A}_i(v)$ are no longer bounded because of the third summand. The output sets extend infinitely far in a finite number of directions given by the subspace in item 3 above. They are also not strictly convex. The projections of the output sets orthogonal to the finite-dimensional subspace are strictly convex. One may visualize the output sets with $\hat{P}_i > 0$ and $\hat{P}_i = 0$ as in figures 4.2.1 and 4.2.2, respectively. In figure 4.2.2 the axis of the infinitely long cylinder is parallel to the subspace spanned by $\{e^{A_i t} x_0\}$ where x_0 is arbitrary.

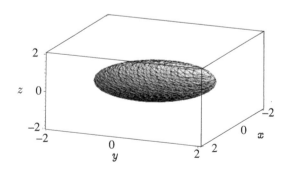

Figure 4.2.1 Output set with $\hat{P}_i > 0$.

The lack of strict convexity can impact on the numerical algorithms used to find separating hyperplanes in section 4.2.5.

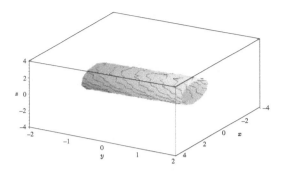

Figure 4.2.2 Output set with $\hat{P}_i = 0$.

4.2.2 The Optimization Formulations

We first recall some ideas from chapter 3 and then set up some needed notation. Let $\mathcal{V}$ denote the set of proper auxiliary signals v and let

$$(\gamma^*)^{-1} = \inf_{v \in \mathcal{V}} \delta(v). \tag{4.2.7}$$

Then γ^* is called the separability index associated with (4.2.1a),(4.2.1b) and norm δ. Note that $\gamma^* = 0$ if there are no L^2 auxiliary signals v which are proper.

For $0 \le \beta \le 1$ we again let

$$\phi_\beta(v) = \inf_{\substack{x_i, \nu_i, y}} \beta S_i(x_i(0), \nu_i) + (1 - \beta) S_i(x_i(0), \nu_i). \tag{4.2.8}$$
$$\text{subject to } (4.2.1a),(4.2.1b)$$

Since we are considering only additive uncertainty here, we can omit the s from the $\phi_\beta(v, s)$ of chapter 3. Recall that $\phi_\beta(v)$ is a nonnegative quadratic function of v which is defined and finite for all v and for $0 \le \beta \le 1$. $\phi_\beta(v)$ is a concave function of β if the set of proper auxiliary signals is not empty, otherwise it is identically zero. Most importantly, we have that v is proper if and only if $\phi_\beta(v) \ge 1$ for some β.

The combined system is

$$\dot{x}_0 = A_0 x_0 + B_0 v + M_0 \nu_0, \tag{4.2.9a}$$
$$y = C_0 x_0 + N_0 \nu_0, \tag{4.2.9b}$$
$$\dot{x}_1 = A_1 x_1 + B_1 v + M_1 \nu_1, \tag{4.2.9c}$$
$$y = C_1 x_1 + N_1 \nu_1. \tag{4.2.9d}$$

Eliminating y, we get that (4.2.9) is

$$\begin{pmatrix} I & 0 \\ 0 & I \\ 0 & 0 \end{pmatrix} \begin{pmatrix} \dot{x}_0 \\ \dot{x}_1 \end{pmatrix} = \begin{pmatrix} A_0 & 0 \\ 0 & A_1 \\ C_0 & -C_1 \end{pmatrix} \begin{pmatrix} x_0 \\ x_1 \end{pmatrix} + \begin{pmatrix} M_0 & 0 \\ 0 & M_1 \\ N_0 & -N_1 \end{pmatrix} \begin{pmatrix} \nu_0 \\ \nu_1 \end{pmatrix} + \begin{pmatrix} B_0 \\ B_1 \\ 0 \end{pmatrix} v.$$
$$\tag{4.2.10}$$

As in section 3.3.5 this descriptor system can be written more succinctly with the obvious correspondences as

$$\dot{x} = Ax + M\nu + Bv, \tag{4.2.11a}$$

$$0 = Cx + N\nu. \tag{4.2.11b}$$

A need not be Hurwitz. The case when A is not Hurwitz is important since it includes the situation where the original system is stable, the failed system is unstable, and we want our detection horizon T to be kept small enough to detect the failure before the instability of the failed system can create problems.

We could use (4.2.11) directly in the algorithms. However, we have found the numerical algorithms to be numerically more robust, if we proceed as follows.

By assumption the N_i are both full row rank. Performing a constant orthogonal change of coordinates U_i on ν_i, we may assume that $N_i = [\bar{N}_i, 0]$ where $\bar{N}_i$ is invertible and $M_i = [\bar{M}_i, \widetilde{M}_i]$. Let $\nu_i = [\bar{\nu}_i^T, \widetilde{\nu}_i^T]^T$ with the same decomposition as N_i. We can proceed using either $\bar{\nu}_i$. We shall use $\bar{\nu}_0$. Then we get that the system (4.2.11) can be written as

$$\begin{pmatrix} I & 0 & 0 \\ 0 & I & 0 \\ 0 & 0 & 0 \end{pmatrix} \begin{pmatrix} \dot{x}_0 \\ \dot{x}_1 \\ \dot{\bar{\nu}}_0 \end{pmatrix} = \begin{pmatrix} A_0 & 0 & \bar{M}_0 \\ 0 & A_1 & 0 \\ C_0 & -C_1 & \bar{N}_0 \end{pmatrix} \begin{pmatrix} x_0 \\ x_1 \\ \bar{\nu}_0 \end{pmatrix}$$
$$+ \begin{pmatrix} \widetilde{M}_0 & 0 & 0 \\ 0 & \bar{M}_1 & \widetilde{M}_1 \\ 0 & -\bar{N}_1 & 0 \end{pmatrix} \begin{pmatrix} \widetilde{\nu}_0 \\ \bar{\nu}_1 \\ \widetilde{\nu}_1 \end{pmatrix} + \begin{pmatrix} B_0 \\ B_1 \\ 0 \end{pmatrix} v. \tag{4.2.12}$$

Multiply the bottom equation by $\bar{N}_0^{-1}$. Again call the new matrices in the last row of (4.2.12) $C_0, C_1, \bar{N}_1$. Then eliminating $\bar{\nu}_0$ we get that the system (4.2.12) is

$$\begin{pmatrix} \dot{x}_0 \\ \dot{x}_1 \end{pmatrix} = \begin{pmatrix} A_0 + \bar{M}_0 C_0 & \bar{M}_0 C_1 \\ 0 & A_1 \end{pmatrix} \begin{pmatrix} x_0 \\ x_1 \end{pmatrix} + \begin{pmatrix} \widetilde{M}_0 & \bar{M}_0 \bar{N}_1 & 0 \\ 0 & \bar{M}_1 & \widetilde{M}_1 \end{pmatrix} \begin{pmatrix} \widetilde{\nu}_0 \\ \bar{\nu}_1 \\ \widetilde{\nu}_1 \end{pmatrix} + \begin{pmatrix} B_0 \\ B_1 \end{pmatrix} v$$
$$\tag{4.2.13}$$

or equivalently

$$\dot{x} = \widetilde{A}x + \widetilde{M}\hat{\nu} + Bv. \tag{4.2.14}$$

We can rewrite $\phi_\beta(v)$ in terms of the simplification (4.2.13). The integral on the right of (4.2.8) can be written

$$\mathcal{J}_v(\beta) = \int_0^T \beta|\widetilde{\nu}_0|^2 + \beta|C_0 x_0 - C_1 x_1 - \bar{N}_1 \bar{\nu}_1|^2 + (1-\beta)|\bar{\nu}_1|^2 + (1-\beta)|\widetilde{\nu}_2|^2 \, dt. \tag{4.2.15}$$

Let

$$\hat{P}_\beta = \mathrm{diag}\{\beta\hat{P}_0, (1-\beta)\hat{P}_1\}.$$

Thus $\phi_\beta(v)$ is the minimum cost in a quadratic cost control problem with cost (4.2.15) and process (4.2.14) and $\hat{\nu}$ is the control. The cost is of the form

$$x(0)^T \hat{P}_\beta x(0) + \int_0^T x^T Q x + x^T H \hat{\nu} + \hat{\nu}^T R \hat{\nu} \, dt$$

and

$$S = \begin{pmatrix} Q & H/2 \\ H^T/2 & R \end{pmatrix} \tag{4.2.16a}$$

is positive semidefinite and positive definite on the $\hat{\nu}$ component. Furthermore, Q, R, H are continuous functions of β since

$$Q = 2\beta \begin{pmatrix} C_0^T C_0 & -C_0^T C_1 \\ -C_1^T C_0 & C_1^T C_1 \end{pmatrix}, \quad H = 4\beta \begin{pmatrix} 0 & -C_0^T \bar{N}_1 & 0 \\ 0 & C_1^T \bar{N}_1 & 0 \end{pmatrix}, \tag{4.2.16b}$$

$$R = 2 \begin{pmatrix} \beta I & 0 & 0 \\ 0 & (1-\beta)I + \beta \bar{N}_1^T \bar{N}_1 & 0 \\ 0 & 0 & (1-\beta)I \end{pmatrix}. \tag{4.2.16c}$$

Later we will also need the matrices

$$\bar{S} = \frac{1}{2}\widetilde{M}R^{-1}H^T, \quad \bar{W} = \frac{1}{4}HR^{-1}H^T, \quad \bar{V} = \widetilde{M}R^{-1}\widetilde{M}^T. \tag{4.2.16d}$$

4.2.3 Computation of the Separability Index and Minimum Norm v

Modern optimization software is capable of handling complex problems with parameters and a variety of constraints. Since γ^* and v can be computed off-line, this means that such approaches become competitive on small and moderate-sized finite intervals. In this section we develop several formulations of the auxiliary signal design problem that are optimization problems. We use the industrial grade optimization package SOCS (sparse optimal control software) [12, 13]. However, any optimal control package could be used that can accommodate state and boundary constraints including inequality constraints. The choice of software has some impact on the numerical experience. We shall comment some more about SOCS in section 4.2.6 just before the computational examples.

There will be four formulations of the auxiliary signal design problem of chapter 3. Two of the formulations described below, formulations (4.2.21) and (4.2.22), are based on

$$(\gamma^*)^2 = \max_{v,\beta} \phi_\beta(v). \tag{4.2.17}$$

They use characterizations of $\phi_\beta(v)$ to reduce the maximum-minimum problem for γ^* to a maximum problem. The other two formulations described below, formulations (4.2.23) and (4.2.25), use the definition of a minimum proper v to characterize the solution of

$$\min_v \delta^2(v) \text{ such that } \phi_\beta(v) \geq 1 \text{ for some } \beta \in [0,1]. \tag{4.2.18}$$

While these formulations are similar for finding the auxiliary signal and in doing the identification in the next section, we shall see that (4.2.18) has advantages when more than two models are involved.

Computational experience has not shown one formulation to be clearly superior to another when there are only two models. Which performs better depends on the length of the interval, the size of the problem, and the dynamics. Formulations 3 and

4 are more useful if there are more than two models, as shown in section 4.4. In all the formulations the auxiliary signal is v. The notation for the other system matrices and variables is from (4.2.20), (4.2.19) and (4.2.13), (4.2.16), and (4.2.14).

Formulations 2 and 4 turn out to be the most generally useful and are the ones that we will use later in this chapter. Since formulations 1 and 3 are included just to illustrate what they look like, we will state formulations 1 and 3 only for the $\hat{P}_\beta = 0, U = 0, W = 0$ case.

We simplify our notation and write (4.2.14) as

$$\dot{x} = Ax + Bv + M\hat{\nu} \qquad (4.2.19)$$

with the system matrices (4.2.13), cost functional (4.2.15), and

$$\hat{\nu} = \begin{pmatrix} \tilde{\nu}_0 \\ \bar{\nu}_1 \\ \tilde{\nu}_1 \end{pmatrix}. \qquad (4.2.20)$$

4.2.3.1 Optimization Problem 1: $(\hat{P}_\beta = 0, U = 0, W = 0)$

The definition of γ^* in (4.2.17) involves a maximization-minimization problem. We shall rewrite this problem by using the necessary conditions to formulate an optimization problem. Boundary inequalities provide a convenient way to implement L^2 restrictions.

Since we will be integrating the equations anyway, and not looking for a feedback solution, we wish to formulate the cost as directly as possible in terms of the solutions of the differential equations.

The first optimization problem is

$$(\gamma^*)^2 = \max_{v,\beta} \left(\varphi(0) - \frac{1}{2}g(0)^T x(0) \right) = - \min_{v,\beta} \left(\frac{1}{2}g(0)^T x(0) - \varphi(0) \right), \qquad (4.2.21a)$$

where

$$\dot{K} = (\bar{S}^T - A^T)K + K(\bar{S} - A) + K\bar{V}K + \bar{W} - Q, \quad K(T) = 0, \quad (4.2.21b)$$

$$\dot{g} = (\bar{S}^T - A^T + K\bar{V})g + KBv, \quad g(T) = 0, \qquad (4.2.21c)$$

$$\dot{x} = (A - \bar{S} - \bar{V}K)x + \bar{V}g + Bv, \qquad (4.2.21d)$$

$$\dot{\varphi} = \frac{1}{2}g^T V g + v^T B^T g, \quad \varphi(T) = 0, \qquad (4.2.21e)$$

$$\dot{\theta} = v^T v, \quad \theta(0) = 0, \theta(T) = 1, \qquad (4.2.21f)$$

$$0 = K(0)x(0) - g(0), \qquad (4.2.21g)$$

$$0 < \beta < 1. \qquad (4.2.21h)$$

Note that we have avoided forming K^{-1} where K is the solution of the Riccati equation (4.2.21b). Equation (4.2.21f) imposes $\|v\| = 1$. The remaining conditions ensure that for a given v, β, that $\phi_\beta(v)$ is the right hand side of (4.2.21a). Once (4.2.21) is solved, we have that v/γ^* is proper of minimum norm.

COMMENT ON COMPUTATION

In solving optimization problem 1, β is treated as a parameter and $v, \hat{\nu}$ are treated as control variables. We have found that it helps the software get a feasible solution on coarse grids if condition (4.2.21h) is replaced by $0.01 \leq \beta \leq 0.99$. We have experimented with a number of different formulations of (4.2.21), including letting $\|v\| \leq 1$, to see if they might make it easier to get feasible solutions on coarse grids. The formulations presented here seemed to work best.

4.2.3.2 Optimization Problem 2

Suppose that $\hat{P}_\beta \geq 0$ and U, W are from (4.2.3). Formulation 1 is based on the usual Riccati theory. However, K is a matrix and this could lead to large systems for complex-higher dimensional problems. If the optimization software proceeds by discretizing the entire problem, then the usual computational advantages of a Riccati formulation over a boundary value formulation are lost. One can work directly from the variational equations. This is especially useful on short intervals. The optimization problem is

$$(\gamma^*)^2 = \max_{v,\beta} z(T) = \min_{v,\beta} -z(T), \qquad (4.2.22a)$$

where

$$\dot{x} = Ax + Bv + M\hat{\nu}, \qquad (4.2.22b)$$

$$\dot{\psi} = F_0\psi + G_0 v, \quad \psi(0) = 0, \qquad (4.2.22c)$$

$$\dot{\lambda} = -Qx - \frac{1}{2}H\hat{\nu} - A^T\lambda, \quad \lambda(0) + \hat{P}_\beta x(0) = 0, \quad \lambda(T) = 0, \quad (4.2.22d)$$

$$\dot{\theta} = v^T v + \psi^T U\psi, \quad \theta(0) = 0, \quad \theta(T) + \psi(T)^T W\psi(T) = 1, \quad (4.2.22e)$$

$$0 = R\hat{\nu} + \left(\frac{1}{2}H^T x + M^T\lambda\right), \qquad (4.2.22f)$$

$$\dot{z} = \frac{1}{2}\left(x^T Qx + x^T H\hat{\nu} + \hat{\nu}^T R\hat{\nu}\right), \quad z(0) = x(0)^T \hat{P}_\beta x(0), \quad (4.2.22g)$$

$$0 < \beta < 1. \qquad (4.2.22h)$$

Then v/γ^* is a proper auxiliary signal of minimum norm δ. Equation (4.2.22f) provides $\hat{\nu}$, which is the worst case noise.

4.2.3.3 Optimization Problem 3 ($\hat{P}_\beta = 0, U = 0, W = 0$)

Another approach is to directly minimize the norm of the proper v. For the system given by (4.2.19) with the cost given by (4.2.15), the separability index as defined by (4.2.7) can be computed using the following optimization problem formulation:

$$(\gamma^*)^{-2} = \min_{v} \|v\|^2 \qquad (4.2.23a)$$

subject to the constraints

$$\dot{K} = (\bar{S}^T - A^T)K + K(S - A) + K\bar{V}K + \bar{W} - Q, \quad K(T) = 0, \text{(4.2.23b)}$$

$$\dot{g} = (\bar{S}^T - A^T + K\bar{V})g + KBv, \quad g(T) = 0, \tag{4.2.23c}$$

$$\dot{x} = (A - S - \bar{V}K)x + \bar{V}g + Bv, \tag{4.2.23d}$$

$$\dot{\varphi} = \frac{1}{2}g^T\bar{V}g + v^T B^T g, \quad \varphi(T) = 0, \tag{4.2.23e}$$

$$0 = K(0)x(0) - g(0), \tag{4.2.23f}$$

$$1 \leq \varphi(0) - \frac{1}{2}x(0)^T g(0), \tag{4.2.23g}$$

$$0 < \beta < 1. \tag{4.2.23h}$$

Equation (4.2.23g) is the requirement that $J_v(\beta) \geq 1$. The optimal (worst case) $\hat{\nu}$ can be obtained from the equation

$$\hat{\nu} = -R^{-1}\left(\frac{1}{2}H^T + M^T K\right)x + R^{-1}M^T g \tag{4.2.24}$$

if one wishes to find the largest perturbation ν_i satisfying the noise bound that goes with v.

4.2.3.4 Optimization Problem 4

We can also use the Euler-Lagrange version of $\phi_\beta(v)$. The norm $\delta(v)$ is given by (4.2.3). Then the optimization problem is

$$(\gamma^*)^{-2} = \min_v \delta^2(v) \tag{4.2.25a}$$

subject to the constraints

$$\dot{x} = Ax + Bv + M\hat{\nu}, \tag{4.2.25b}$$

$$\dot{\psi} = F_0\psi + G_0 v, \quad \psi(0) = 0, \tag{4.2.25c}$$

$$\dot{\lambda} = -Qx - \frac{1}{2}H\hat{\nu} - A^T\lambda, \quad \lambda(0) + \hat{P}_\beta x(0) = 0, \ \lambda(T) = 0, \tag{4.2.25d}$$

$$0 = R\hat{\nu} + \left(\frac{1}{2}H^T x + M^T\lambda\right), \tag{4.2.25e}$$

$$\dot{z} = \frac{1}{2}\left(x^T Q x + x^T H\hat{\nu} + \hat{\nu}^T R\hat{\nu}\right), \quad z(0) = x(0)^T \hat{P}_\beta x(0), \tag{4.2.25f}$$

$$0 < \beta < 1, \tag{4.2.25g}$$

$$z(T) \geq 1. \tag{4.2.25h}$$

4.2.4 Model Identification

We now turn to the problem of model identification. That is, given that we are applying a proper auxiliary signal, how do we decide which model is correct? This was discussed in section 3.3. Here we use a hyperplane test. One technique for computing a hyperplane test was given in theorem 3.3.5. Here we develop an alternative algorithm. The development assumes that v is proper but not necessarily the minimum proper auxiliary signal. This will be useful when we consider more than two models.

Suppose that we have a proper auxiliary signal v that has been computed by one of the approaches in the previous sections. Let $\bar{\mathcal{A}}_i(v)$ be the set of outputs from model i with noise satisfying $\mathcal{S}_i(x_i(0), \nu_i) \leq 1$ instead of $\mathcal{S}_i(x_i(0), \nu_i) < 1$. Since the N_i are full row rank we may conclude that the $\mathcal{A}_i(v)$ are open convex sets whose closures in L^2 are the $\bar{\mathcal{A}}_i(v)$. Furthermore, we have by the properness of v that $\mathcal{A}_0(v) \cap \mathcal{A}_1(v) = \emptyset$. Thus we can separate the $\bar{\mathcal{A}}_i(v)$ with a hyperplane. That is, there are functions $a, \bar{y}$ in L^2 such that if we define

$$\psi(y) = \langle a, y - \bar{y} \rangle = \int_0^T a(t)^T (y(t) - \bar{y}(t)) dt$$

we have that ψ is nonnegative on one $\mathcal{A}_i(v)$ and nonpositive on the other $\mathcal{A}_i(v)$. $\bar{y}$ is unique if one of the $\mathcal{A}_i(v)$ is strictly convex and v is the minimum proper auxiliary signal since then $\bar{y} \in \bar{\mathcal{A}}_0(v) \cap \bar{\mathcal{A}}_1(v)$.

Note that, when v is a minimum proper auxiliary signal and the method of this chapter is used, then the optimization problem that solves for v is formulated in such a way that the output $\bar{y}$ is not immediately available. The output $\bar{y}$ must be reconstructed from the output of the optimization problem. Suppose formulation (4.2.22) is used to find γ^*. The optimization software will return the solution vectors for all system variables $(x, \tilde{v}, \hat{\nu}, \lambda, \psi, z)$ and parameters β. For this problem, $\tilde{v}/\gamma^*$ is $\hat{v}$. To reconstruct $\bar{y}$, note that $\hat{\nu}$ gives $\bar{\nu}_1, \tilde{\nu}_1$ and $\nu_1 = \begin{pmatrix} \bar{\nu}_1 \\ \tilde{\nu}_1 \end{pmatrix}$. Thus $\bar{y}$ can be computed by using the appropriate elements of x and ν from the optimization software in the equation

$$\bar{y} = C_1 x_1 + N_1 \nu_1. \tag{4.2.26}$$

Since the identification will be done in the presence of numerical and system errors one will usually want to apply δv with $\delta > 1$.

The equations (4.2.9) are linear. Thus if v is the minimum norm auxiliary signal we have that δv is the minimum norm auxiliary signal if the original noise bound $\mathcal{S}(x_i(0), \nu_i) < 1$ is relaxed to $\mathcal{S}(x_i(0), \nu_i) < \delta$ and $\delta \bar{y}$ is the intersection of the closures of the new larger output sets which contain $\mathcal{A}_i(\delta v)$. The normal of the new separating hyperplane is still $a(t)$. Accordingly, we may use the identification test

$$\psi_\delta(y) = \int_0^T a(t)^T (y(t) - \delta \bar{y}(t)) dt = \int_0^T a(t)^T y(t) dt - \delta \sigma,$$

which properly separates $\mathcal{A}_0(\delta v)$ and $\mathcal{A}_1(\delta v)$ for $\delta > 1$. Note that σ, which is $\int_0^T a(t)^T \bar{y}(t) dt$, may be precomputed so that in practice $\bar{y}$ does not have to be stored. Only $a(t)$ must be stored.

4.2.5 Computation of Separating Hyperplane

In order to apply the hyperplane test we need $\bar{y}(t)$ and the normal $a(t)$. If a minimum energy auxiliary signal has been computed for two models by the technique of either chapter 3 or this section, then as noted $\bar{y}$ is already available. In the algorithm that follows we shall assume only that we have a proper v and point out where $\bar{y}$ can be used if it is also available.

The approach of this section is to slightly pull back the two output sets and then find the shortest distance between them. Note that this pulling back also makes the numerical calculation more robust because, even if $\bar{y}$ is unique theoretically, the discrete output sets might well overlap and have a nonunique value which would create numerical difficulties when finding a. The normal produced by differencing the closest points in the closures of the two sets will approximate the true normal to the separating hyperplane between the sets computed when the noise vectors are fully considered.

Separating Hyperplane Algorithm

This algorithm computes a vector $\bar{y}$ and an estimate of the normal to the separating hyperplane for the sets $\mathcal{A}_0(v), \mathcal{A}_1(v)$.

1. Determine a proper auxiliary signal v.

2. Reconstruct $\bar{y}$ from the output using (4.2.26) if v was found using the method of this chapter and is the minimum proper auxiliary signal. Otherwise, see step 5 below.

3. If v is known to not be the minimum proper auxiliary signal, then one can define the parameter ϵ to be 1. Otherwise, choose ϵ to be $0 < \epsilon < 1$ and close to 1. Solve the optimization problem

$$\min \|y_0 - y_1\|^2 \tag{4.2.27}$$

subject to the constraints

$$\dot{x}_0 = A_0 x_0 + B_0 v + M_0 \epsilon \nu_0, \tag{4.2.28a}$$
$$y_0 = C_0 x_0 + N_0 \epsilon \nu_0, \tag{4.2.28b}$$
$$\dot{x}_1 = A_1 x_1 + B_1 v + M_1 \epsilon \nu_1, \tag{4.2.28c}$$
$$y_1 = C_1 x_1 + N_1 \epsilon \nu_1, \tag{4.2.28d}$$
$$\dot{w}_1 = \nu_0^T \nu_0, \quad w_1(0) = 0, \quad w_1(T) \le 1, \tag{4.2.28e}$$
$$\dot{w}_2 = \nu_1^T \nu_1, \quad w_2(0) = 0, \quad w_2(T) \le 1. \tag{4.2.28f}$$

4. Compute $a(t)$, the normal to the separating hyperplane, as

$$a = \frac{y_0 - y_1}{\|y_0 - y_1\|}.$$

5. If $\bar{y}$ is not known, then let $\bar{y} = \frac{1}{2}(y_0 + y_1)$ where y_0, y_1 are from the solution of (4.2.27), (4.2.28).

6. Let $\psi(z) = \langle a, z - \bar{y} \rangle$ be the test function.

Note that if v is the minimum proper auxiliary signal, then as ϵ approaches 1, the computed normal approaches the true normal, but the numerical error increases once ϵ is very small due to division by small numbers. Computational experience to date suggests that the examples at the end of this section are typical in that $a(t)$ is not very sensitive to ϵ as long as ϵ is not too far from 1. Finally, note that if $\mathcal{A}_{\epsilon,i}(v)$ is the output from model i using input v and $\epsilon x_i(0)$, $\epsilon \nu_i$, then $\mathcal{A}_{1,i}(v) = \mathcal{A}_i(v)$ and $\mathcal{A}_{\epsilon,i}(v) = \epsilon \mathcal{A}_i(\frac{1}{\epsilon} v)$.

The separating hyperplane algorithm given here computes only an approximation to the normal $a(t)$. The next result addresses the accuracy of the resulting hyperplane test.

Theorem 4.2.1 *Let $a_\epsilon(t)$ be the normal from the separating hyperplane algorithm using $0 < \epsilon < 1$ and let $\bar{y}_{0,\epsilon}, \bar{y}_{1,\epsilon}$ be the values of y_0, y_1 that give the minimum distance. Let $\bar{y}_\epsilon = \dfrac{1}{2}(\bar{y}_{0,\epsilon} + \bar{y}_{1,\epsilon})$. Let the hyperplane test be*

$$\psi_\epsilon(s) = \langle a_\epsilon(t), s - \bar{y}_\epsilon \rangle. \tag{4.2.29}$$

Let $d(\epsilon) = \|\bar{y}_{0,\epsilon} - \bar{y}_{1,\epsilon}\|/2$. Then there is a constant K so that $d(\epsilon) \le K(1 - \epsilon)$ and if $s \in (\mathcal{A}_{\epsilon,0}(v) \cup \mathcal{A}_{\epsilon,1}(v))$, then

$$\psi_\epsilon(s) \ge d(\epsilon) \implies s \in \mathcal{A}_{\epsilon,0}(v), \tag{4.2.30}$$

$$\psi_\epsilon(s) \le -d(\epsilon) \implies s \in \mathcal{A}_{\epsilon,1}(v). \tag{4.2.31}$$

Proof. (4.2.30) and (4.2.31) follow from noting that $\psi_\epsilon(s) = d(\epsilon)$ is the supporting hyperplane of $\mathcal{A}_{\epsilon,0}(v)$ at $\bar{y}_{0,\epsilon}$ while $\psi_\epsilon(s) = -d(\epsilon)$ is a supporting hyperplane of $\mathcal{A}_{\epsilon,1}(v)$ at $\bar{y}_{1,\epsilon}$. Then K can be taken as $\|C_0\mathcal{L}_0 M_0 + N_0\| + \|C_1\mathcal{L}_1 M_1 + N_1\|$. $\square$

The tests are often much better than indicated by theorem 4.2.1, but this statement allows for highly skewed convex sets.

The next section gives a number of computational examples. They illustrate ideas from this and earlier chapters. But before presenting the examples, we will comment in some more detail on the software used and on some general issues concerning numerical optimization.

4.2.6 Comment on Software & Numerics

The problem formulations and approach of this chapter are numerical and their solution requires high-quality optimization software. There are a variety of good optimization software packages available. In order to handle the problems we have described in this chapter the software needs to be able to accept optimal control problems with equality and inequality constraints both inside the time interval and at the end points. As mentioned earlier, the particular package we have used is called SOCS and was developed at the Boeing Corporation [9, 12, 13, 14].

There are a number of numerical approaches to doing optimal control. One of the most popular is the direct transcription approach. In this approach all differential equations are discretized, as are all costs that involve integrals. The resulting large nonlinear programming (NLP) problem is then passed to an optimization package. SOCS is a direct transcription code. The user supplies the equations and constraints.

The discretization is performed within SOCS. The default discretization in SOCS is a fourth-order Runge-Kutta method based on Hermite-Simpson interpolation. Initially, SOCS starts with a coarse time grid but then iteratively refines the grid using sophisticated mesh refinement strategies [10, 11]. In discussing the computational examples we shall make some comments on our computational experience. While all of this experience is based on SOCS, similar comments should hold when using any state of the art direct transcription optimal control code.

A few general observations should be made. First, the minimum proper auxiliary signal is never unique. For linear problems with no inputs other than v and with additive noise, we have that $-v$ is proper if v is. In general, there are always at least two minimum proper v. This means that there will be a saddle point between them. Thus starting with an initial guess of zero may slow down convergence since the method would be near a saddle point.

The second observation is that if SOCS has numerical problems, they usually occur on the initial coarse grid when trying to find a feasible solution of the NLP problem. On coarse grids, the discretization may not be resolving the dynamics well, and if the interval is small the auxiliary signal may be large. Accordingly, one is better off starting with a large initial guess since that is more likely to be in the feasible region.

Finally, there is the question of whether one is finding the true minimum. SOCS does not check whether it is finding the true minimum. Its philosophy is that if it has solved the discrete approximate problem well, and if the discretization is capturing the dynamics well, then the answer is acceptable. Since it was developed to solve complex optimal control problems in the aerospace industry, this is reasonable. What is important is whether you are getting close to the optimal cost and that the system will act as you expect with the control, and not whether you are actually close to the true optimal control. Within our context, the same applies. The v found will be proper. If a good quality proper auxiliary signal is found it can be used. Having said that, we have not seen any problems where the method failed to find an optimum. In fact, we deliberately created some problems with no weight on the initial state for which the output sets had a parallel face. This resulted in the normal of the separating hyperplane being unique but not its end points y_0, y_1. The code still converged.

A large number of examples and test problems have been solved. The interested reader is referred to [43] for a detailed presentation of these tests. Here we make a few summary comments about these tests before giving specific examples.

For low-dimensional models none of the four formulations was clearly superior. The CPU times varied with T in a nonmonotonic manner. Figure 4.2.3 gives the plot of γ^* for a typical example. Notice that for this example γ^* has started to level off by $T = 10$. As with all problems, for smaller T, the quantity $(\gamma^*)^{-1}$, which is the norm of the minimum energy auxiliary signal, gets very large. For this example, this is already significant by $T = 1$. The small value of γ^* sometimes led to premature termination for small T. This was easily corrected by rescaling the cost functions by 10^6. The values of γ^*, β computed by the different formulations were essentially equal. The γ^* values agreed to eight places and the β values to five places. We varied T from 1 to 100 in the examples. The resulting β values varied from 0.24 to 0.51 but values between 0.40 and 0.50 were the most common.

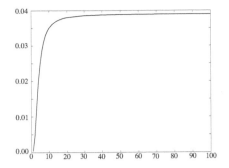

Figure 4.2.3 Plot of γ^* for a typical example.

4.2.7 Two-Model Examples

Our first example is an academic example that illustrates several points.

Example 4.2.1 *In this example, the eigenvalues of A_i change from $\pm 3i$ in model 0 to $\pm 2i$ in model 1. Let $\hat{P} = 0$ and $U = 0$, $W = 0$. The two models are*

$$\dot{x}_0 = \begin{pmatrix} 0 & 1 \\ -9 & 0 \end{pmatrix} x_0 + \begin{pmatrix} 0 \\ 1 \end{pmatrix} v + \begin{pmatrix} 1 & 0 & 0 \\ 0 & 1 & 0 \end{pmatrix} \nu_0, \qquad (4.2.32a)$$

$$y = \begin{pmatrix} 1 & 0 & 0 \end{pmatrix} x_0 + \begin{pmatrix} 0 & 0 & 1 \end{pmatrix} \nu_0, \qquad (4.2.32b)$$

$$\dot{x}_1 = \begin{pmatrix} 0 & 1 \\ -4 & 0 \end{pmatrix} x_1 + \begin{pmatrix} 0 \\ 1 \end{pmatrix} v + \begin{pmatrix} 1 & 0 & 0 \\ 0 & 1 & 0 \end{pmatrix} \nu_1, \qquad (4.2.32c)$$

$$y = \begin{pmatrix} 1 & 0 & 0 \end{pmatrix} x_1 + \begin{pmatrix} 0 & 0 & 1 \end{pmatrix} \nu_1. \qquad (4.2.32d)$$

When $T = 1$ and using optimization formulation 2, example 4.2.1 was one of the problems discussed previously that required multiplying the cost by 10^6 and using a nonzero initial guess of $v = 1$ to get convergence. Figure 4.2.4 shows v for $T = 1, 20$.

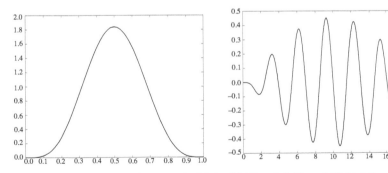

Figure 4.2.4 v for example 4.2.1 with $T = 1$ (left) and 20 (right).

If we apply the separating hyperplane algorithm of section 4.2.5, we get $\bar{y}$ and a for example 4.2.1 as shown in figure 4.2.5. The approximate normal $a_\epsilon(t)$ to the

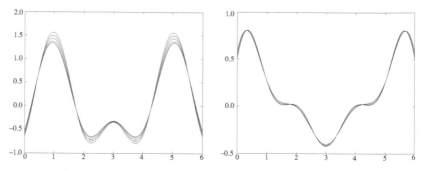

Figure 4.2.5 $\bar{y}$ and $a(t)$ for example 4.2.1; $\epsilon = 0.3, 0.5, 0.7, 0.9$.

hyperplane is sometimes independent of the parameter ϵ. Here it is not. Figure 4.2.5 is typical in that the variation in a_ϵ and $\bar{y}$ for ϵ close to 1 is small. It should be noted that we have used fairly stringent tolerances (10^{-6}) when using the optimization software. One can still get good approximations to the auxiliary signal merely by requesting less accurate numerical solutions. The symmetrical shape in figure 4.2.5 is not a general characteristic of these problems if $P \neq 0$ or if the coefficients are time varying.

If the two models have some common modes which appear in the output and $\hat{P}_\beta = 0$, then it is possible for the two output sets to have a parallel face. If the parallel face includes the closest points between the two sides, then a is unique but $\bar{y}$ is not unique. This could be expected to pose numerical problems for the optimization software. We tried to solve such an example. The optimization software stopped when it was close to meeting the termination tolerances with an error message warning of dependent constraints. However, it had already computed a high enough quality v and $\bar{y}$. We were then able to compute a normal a.

Example 4.2.2 *The equalized and linearized model of a single engine F-16 aircraft [86] is represented by*

$$\dot{x}_0 = \begin{pmatrix} -0.1689 & 0.0759 & -0.9952 \\ -26.859 & -2.5472 & 0.0689 \\ 9.3603 & -0.1773 & -2.4792 \end{pmatrix} x_0 + \begin{pmatrix} 0 & 0 \\ 1 & 0 \\ 0 & 1 \end{pmatrix} v + \begin{pmatrix} 0 & 0 & 0 & 0 & 1 \\ 0 & 0 & 0 & 1 & 0 \\ 0 & 0 & 1 & 0 & 0 \end{pmatrix} \nu_0,$$

$$(4.2.33\text{a})$$

$$y = \begin{pmatrix} 1 & 0 & 0 \\ 0 & 0.9971 & 0.0755 \end{pmatrix} x_0 + \begin{pmatrix} 0 & 1 & 0 & 0 & 0 \\ 1 & 0 & 0 & 0 & 0 \end{pmatrix} \nu_0. \qquad (4.2.33\text{b})$$

A failure simulating an electrical interruption to a flight control computer's input channels may be represented as

$$\dot{x}_1 = \begin{pmatrix} 1 & 1 & 0 \\ 0 & 1 & 1 \\ 0 & 0 & 1 \end{pmatrix} x_1 + \begin{pmatrix} 0 & 0 \\ 1 & 0 \\ 0 & 1 \end{pmatrix} v + \begin{pmatrix} 0 & 0 & 0 & 0 & 1 \\ 0 & 0 & 0 & 1 & 0 \\ 0 & 0 & 1 & 0 & 0 \end{pmatrix} \nu_1, \quad (4.2.33\text{c})$$

$$y = \begin{pmatrix} 1 & 0 & 0 \\ 0 & 0.9971 & 0.0755 \end{pmatrix} x_1 + \begin{pmatrix} 0 & 1 & 0 & 0 & 0 \\ 1 & 0 & 0 & 0 & 0 \end{pmatrix} \nu_1. \qquad (4.2.33\text{d})$$

We take $\hat{P} = 0, W = 0, U = 0.$

Here the three states are side slip, roll rate, and yaw rate. The control is $v = \begin{pmatrix} v_1 \\ v_2 \end{pmatrix}$, where v_1 is the side slip acceleration command (rudder input) and v_2 is the stability axis roll acceleration command (stick input). We take $\hat{P}_\beta = 0$. Figures 4.2.6 and 4.2.7 give $v, \bar{y}_\epsilon(t)$, and $a_\epsilon(t)$ for this problem on the interval $[0, 1]$.

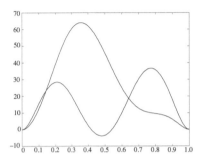

Figure 4.2.6 Minimum proper v for example 4.3.1 with $T = 1$.

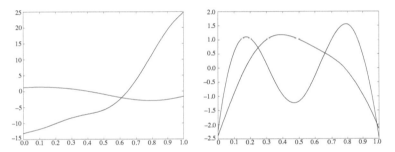

Figure 4.2.7 $\bar{y}_\epsilon(t)$ and $a_\epsilon(t)$ for example 4.3.1 using $\epsilon = 0.7$ and $T = 1$.

4.2.8 Example Showing Effect of $\hat{P}_i$

It is interesting to observe how the weight on the initial state affects the shape of the auxiliary signal. If the two models are stable, then the larger the weight, the smaller the transient response. It then becomes advantageous to apply more of the auxiliary signal earlier in the test period. This is illustrated by the next example.

Let $\mathcal{A}_i(v, P_i)$ be the output set with auxiliary signal v and weight P_i in (4.2.2). As P_i is decreased the allowable noise set increases monotonically. Thus $P \leq Q$, that is, $Q - P$ is positive semidefinite, implies that $\mathcal{A}_i(v, Q) \subset \mathcal{A}_i(v, P)$. The norm of the minimum norm auxiliary signal varies monotonically from a nonzero lower bound to a nonzero upper bound as the P_i decrease to zero. We also get that if v is proper for some P_i, then it will also be proper for any larger P_i.

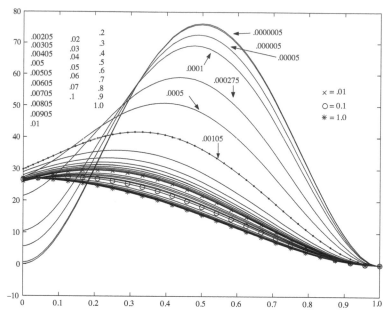

Figure 4.2.8 Minimum proper v for (4.2.34) on $[0, 1]$ for $0 \leq P \leq 1$.

Example 4.2.3 *Let* $P = P_0 = P_1$ *and* $U = 0$, $W = 0$. *Let the two models be*

$$\dot{x}_0 = -2x_0 + v + \begin{pmatrix} 1 & 0 \end{pmatrix} \nu_0, \qquad (4.2.34a)$$

$$y = x_0 + \begin{pmatrix} 0 & 1 \end{pmatrix} \nu_0, \qquad (4.2.34b)$$

$$\dot{x}_1 = -x_1 + v + \begin{pmatrix} 1 & 0 \end{pmatrix} \nu_1, \qquad (4.2.34c)$$

$$y = x_1 + \begin{pmatrix} 0 & 1 \end{pmatrix} \nu_1. \qquad (4.2.34d)$$

Figure 4.2.8 shows the minimum norm proper auxiliary signal for several values of P for $0 \leq P \leq 1$ and $T = 1$. The values of P are indicated on the graph. It is interesting to note that the transition happens very quickly, so that even a small amount of P_i gives a minimum proper auxiliary signal that has energy almost down to the value with a large P_i.

4.3 GENERAL M-MODEL CASE

We now consider the situation where there are m models of the form (4.2.1). One may think of model 0 as normal operation and the other $m - 1$ models as models of various failures. The outputs are measured, and the inputs (auxiliary signal) can be chosen. We now seek an optimal v that can be used to simultaneously test for all possible failure models. Let v denote an auxiliary signal and let $\mathcal{A}_i(v)$ represent the set of possible outputs y associated with this input, if model i were the correct model. Perfect m-model identification means that

$$\mathcal{A}_i(v) \cap \mathcal{A}_j(v) = \emptyset \quad \text{for} \quad i \neq j, \quad 0 \leq i, j \leq m - 1. \tag{4.3.1}$$

An auxiliary signal which gives (4.3.1) is called a proper m-model auxiliary signal.

One major advantage of the optimization formulations given in this chapter is that the technique directly extends to more than two models. We show how to solve a general m-model problem by describing the three-model case. Suppose that we want a three-model proper auxiliary signal on $[0, 3]$. One way to do this would be to compute a minimum energy auxiliary signal for each of the three individual comparisons on $[0, 1]$ and then apply the auxiliary signals sequentially. We shall refer to this as performing a sequential three-model test on $[0, T]$. The optimization approach works equally well no matter what the weight is on the initial conditions. However, we shall take $\hat{P} = 0$ in the examples since then there is a more transparent relationship between the sizes of perturbations on $[0, 1], [1, 2], [2, 3]$ and perturbations on $[0, 3]$. Note that if v is proper for $\hat{P} = 0$, then it is proper for any $\hat{P} \geq 0$ since increasing $\hat{P}$ reduces the size of the output sets.

Computational examples show that multimodel auxiliary signals can frequently be found that are able to perform three or more tests simultaneously on an interval $[0, T]$ for the same energy as, or only a little more than, that required to do one test on that same interval.

Suppose then that we have three models $i = 0, 1, 2$ in the form of (4.2.1). Let $\phi_\beta^{ij}(v)$ be the $\phi_\beta(v)$ function if model i is being compared to model j with $i \neq j$. Suppose that we wish to compute the minimum energy auxiliary signal v which can be used to simultaneously determine which of the three models is correct. This auxiliary signal is given as the solution of the optimization problem

$$\min \delta(v)^2 \tag{4.3.2a}$$

subject to

$$\phi_{\beta_1}^{01}(v) \geq 1 \text{ for some } \beta_1, \tag{4.3.2b}$$

$$\phi_{\beta_2}^{02}(v) \geq 1 \text{ for some } \beta_2, \tag{4.3.2c}$$

$$\phi_{\beta_3}^{12}(v) \geq 1 \text{ for some } \beta_3. \tag{4.3.2d}$$

If there are m models and we wish to know which model is correct, then we must do $m(m - 1)/2$ comparisons, and there will be that many inequality constraints.

We can use optimization formulation 3 or 4 of section 4.2. If formulation 4 is used, the constraints (4.3.2b)–(4.3.2d) become

$$\dot{\bar{x}}_1 = A^1 \bar{x} + B^1 v + M^1 \hat{v}^1, \quad \lambda_1(0) + \hat{P}_{\beta_1} \bar{x}_1(0) = 0, \tag{4.3.3a}$$

$$\dot{\lambda}_1 = -Q^1 \bar{x}_1 - \frac{1}{2} H^1 \hat{v}^1 - (A^1)^T \lambda_1, \quad \lambda_1(T) = 0, \tag{4.3.3b}$$

$$0 = R^1 \hat{v}^1 + \left(\frac{1}{2} (H^1)^T \bar{x}_1 + M_1^T \lambda_1 \right), \tag{4.3.3c}$$

$$\dot{z}_1 = \frac{1}{2}\left(\bar{x}_1^T Q^1 \bar{x}_1 + \bar{x}_1^T H^1 \hat{\nu}^1 + (\hat{\nu}^1)^T R^1 \hat{\nu}^1\right), \quad z_1(0) = 0, \qquad (4.3.3\text{d})$$

$$0 < \beta_1 < 1, \qquad (4.3.3\text{e})$$

$$1 \le z_1(T) + \bar{x}_1(0)^T \hat{P}_{\beta_1} \bar{x}_1(0), \qquad (4.3.3\text{f})$$

$$\dot{\bar{x}}_2 = A^2 \bar{x}_2 + B^2 v + M^2 \hat{\nu}^2, \quad \lambda_2(0) + \hat{P}_{\beta_2} \bar{x}_2(0) = 0, \qquad (4.3.3\text{g})$$

$$\dot{\lambda}_2 = -Q^2 \bar{x}_2 - \frac{1}{2} H^2 \hat{\nu}^2 - (A^2)^T \lambda_2, \quad \lambda_2(T) = 0, \qquad (4.3.3\text{h})$$

$$0 = R^2 \hat{\nu}^2 + \left(\frac{1}{2}(H^2)^T \bar{x}_2 + (M^2)^T \lambda_2\right), \qquad (4.3.3\text{i})$$

$$\dot{z}_2 = \frac{1}{2}\left(\bar{x}_2^T Q^2 \bar{x}_2 + \bar{x}_2^T H^2 \hat{\nu}^2 + (\hat{\nu}^2)^T R^2 \hat{\nu}^2\right), \quad z_2(0) = 0, \qquad (4.3.3\text{j})$$

$$0 < \beta_2 < 1, \qquad (4.3.3\text{k})$$

$$1 \le z_2(T) + \bar{x}_2(0)^T \hat{P}_{\beta_2} \bar{x}_2(0), \qquad (4.3.3\text{l})$$

$$\dot{\bar{x}}_3 = A^3 \bar{x}_3 + B^3 v + M^3 \hat{\nu}^3, \quad \lambda_3(0) + \hat{P}_{\beta_3} \bar{x}_3(0) = 0, \qquad (4.3.3\text{m})$$

$$\dot{\lambda}_3 = -Q^3 \bar{x}_3 - \frac{1}{2} H^3 \hat{\nu}^3 - (A^3)^T \lambda_3, \quad \lambda_3(T) = 0, \qquad (4.3.3\text{n})$$

$$0 = R^3 \hat{\nu}^3 + \left(\frac{1}{2}(H^3)^T \bar{x}_3 + (M^3)^T \lambda_3\right), \qquad (4.3.3\text{o})$$

$$\dot{z}_3 = \frac{1}{2}\left(\bar{x}_3^T Q^3 \bar{x}_3 + \bar{x}_3^T H^3 \hat{\nu}^3 + (\hat{\nu}^3)^T R^3 \hat{\nu}^3\right), \quad z_3(0) = 0, \qquad (4.3.3\text{p})$$

$$0 < \beta_3 < 1, \qquad (4.3.3\text{q})$$

$$1 \le z_3(T) + \bar{x}_3(0)^T \hat{P}_{\beta_3} \bar{x}_3(0), \qquad (4.3.3\text{r})$$

$$\dot{\psi} = F_0 \psi + G_0 v, \quad \psi(0) = 0. \qquad (4.3.3\text{s})$$

Here the superscript denotes the comparison so that 1 is the comparison between models 0 and 1. Note that there are three independent β's. Similarly the noise vectors $\hat{\nu}^i$ are independent. The size of the problem increases with both the size and the number of models. Usually the greatest difficulty with the larger, more complex optimization problems when using a direct transcription code is in finding feasible solutions on coarse grids. For multimodel problems one can use information from the simpler-two model problems to find feasible solutions. One could, for example, use a proper sequential three-model auxiliary signal as an initial guess.

Given that the auxiliary signal v is found, one sets up the three separating hyperplane tests as before. Note that a v that is a minimum proper auxiliary signal for an m-model comparison with $m > 2$ is not necessarily a minimum proper auxiliary signal for any of the individual two-model comparisons.

There are a number of variations on this problem. For example, it might be desirable to distinguish model 1 from models 0 and 2, but the user might not care about distinguishing between models 0 and 2. Then one solves the problem

$$\min \delta(v)^2, \qquad (4.3.4\text{a})$$

$$\phi_{\beta_1}^{01}(v) \ge 1 \text{ for some } \beta_1, \qquad (4.3.4\text{b})$$

$$\phi_{\beta_3}^{12}(v) \ge 1 \text{ for some } \beta_3 \qquad (4.3.4\text{c})$$

and sets up two hyperplane tests.

Let $\mathcal{P}_{ij}$ be the set of v that are not proper when applied to models i and j. Then $\mathcal{P}_{ij}$ is a convex set including the origin. Doing more than one comparison with the same minimum proper auxiliary signal means finding the smallest v in that norm which is not in the union of several convex sets.

For example, (4.3.2) could be written as

$$\min \delta(v)^2, \tag{4.3.5a}$$

$$v \notin (\mathcal{P}_{01} \cup \mathcal{P}_{02} \cup \mathcal{P}_{12}). \tag{4.3.5b}$$

This has several consequences for any computational algorithm. One is that for three or more comparisons it is possible that the minimum will occur at a cusp, that is, a point on the boundary of the union of the $\mathcal{P}_{i,j}$ where the surface is not differentiable. Consider figure 4.3.1, which shows the union of three ellipsoids. Depending on how the ellipsoids are oriented, the point on the surface of the union that is closest to the origin will lie either on one of the faces or at an intersection. Some experimentation shows that both cases occur frequently. If the minimum lies at an intersection of the surfaces, this may slow down convergence. Another consequence is that there are possibly several local minima so that the optimization software may not find the global minimum. This will not be significant in many applications since the local minimum will often still be a good auxiliary signal. Some computational examples are given in [43].

Figure 4.3.1 Union of three ellipsoids.

It is theoretically possible that a local minimum will actually lie on a continuum of minima if the convex sets are correctly aligned. This will again have the effect of slowing convergence as the numerical method approaches a minimum. However, again the approximations will often still be good auxiliary signals.

The above discussion illustrates an important feature of the approach of this chapter. With several models the actual infinite-dimensional surface in v space being minimized over can be quite complex. However, we have shown how a proper m-model auxiliary signal can always be constructed using a sequential signal. Our algorithm can take this signal as a starting point and then produce a proper m-model signal of smaller norm, usually much smaller norm. As long as the optimization algorithm being used is one that maintains feasibility during the optimization process, and SOCS is such an algorithm, the result is an auxiliary signal that is better than the initial one. Given this v, one may then construct hyperplane tests. It is important to note that all the separating hyperplane tests can be carried out in parallel on separate processors so that the on-line computational effort is essentially independent of the number of models.

4.3.1 Three-Model Example

Example 4.3.1 *Suppose that we have the following three models with $T = 3$ and $\hat{P}_\beta = 0, U = 0, W = 0$:*

$$\dot{x}_0 = \begin{pmatrix} -1 & 2 \\ -1 & 3 \end{pmatrix} x_0 + \begin{pmatrix} 0 \\ 1 \end{pmatrix} v + \begin{pmatrix} 0 & 1 & 0 \\ 0 & 0 & 1 \end{pmatrix} \nu_0, \tag{4.3.6a}$$

$$y = \begin{pmatrix} 1 & 1 \end{pmatrix} x_0 + \begin{pmatrix} 1 & 0 & 0 \end{pmatrix} \nu_0, \tag{4.3.6b}$$

$$\dot{x}_1 = \begin{pmatrix} -10 & 2 \\ -1 & 3 \end{pmatrix} x_1 + \begin{pmatrix} 0 \\ 1 \end{pmatrix} v + \begin{pmatrix} 0 & 1 & 0 \\ 0 & 0 & 1 \end{pmatrix} \nu_1, \tag{4.3.6c}$$

$$y = \begin{pmatrix} 1 & 1 \end{pmatrix} x_1 + \begin{pmatrix} 1 & 0 & 0 \end{pmatrix} \nu_1, \tag{4.3.6d}$$

$$\dot{x}_2 = \begin{pmatrix} 1 & 2 \\ -1 & 0 \end{pmatrix} x_2 + \begin{pmatrix} 0 \\ 1 \end{pmatrix} v + \begin{pmatrix} 0 & 1 & 0 \\ 0 & 0 & 1 \end{pmatrix} \nu_2, \tag{4.3.6e}$$

$$y = \begin{pmatrix} 1 & 1 \end{pmatrix} x_2 + \begin{pmatrix} 1 & 0 & 0 \end{pmatrix} \nu_2. \tag{4.3.6f}$$

On the left hand side of figure 4.3.2 is a comparison of v constructed using the sequential method to that of the simultaneous method. The dashed line graph oscillating minimally around the horizontal axis is the simultaneous auxiliary signal. The simultaneous method produces the solid line v, which is of much higher energy. On the right side of figure 4.3.2 is a comparison of v from the sequential method (dashed line) with the three v obtained by solving the pairwise problems independently (solid lines) on $[0, 3]$. The simultaneous v is different from the pairwise v. Figure 4.3.2 shows that we can often simultaneously test multiple failure models with only slightly more energy than required for a single failure model.

Figure 4.3.3 gives the $\bar{y}_\epsilon(t)$ and $a_\epsilon(t)$ in the hyperplane tests for the sequential method. Figure 4.3.4 gives the same signals for the simultaneous method. Note that in the simultaneous approach there is one v for all three comparisons, but each comparison needs its own hyperplane test, so that there are three $\bar{y}_\epsilon(t)$ and three $a_\epsilon(t)$.

4.4 EARLY DETECTION

In the approach being presented in this chapter, the only computation that cannot be done off-line is the hyperplane test. It is often possible to reject or accept a model before the end of the test period. We shall call this early detection. One way to often make an early decision was described earlier in section 2.3.3. However, that approach requires working with the state vectors. Here we show one way to modify the hyperplane approach to permit early decision making using only the usually much smaller output sets. This requires considerably less computational storage and effort than the other approach.

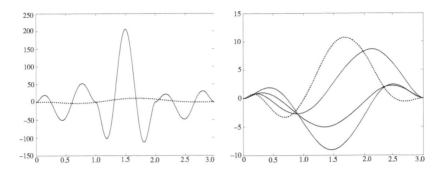

Figure 4.3.2 v for example 4.3.1: sequential versus simultaneous (left), and three full interval two-model versus simultaneous (right).

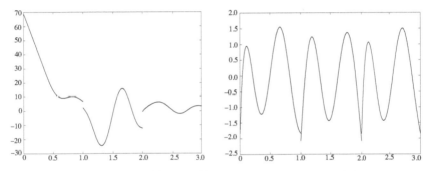

Figure 4.3.3 $\bar{y}_\epsilon(t)$ and $a_\epsilon(t)$ for example 4.3.1: sequential.

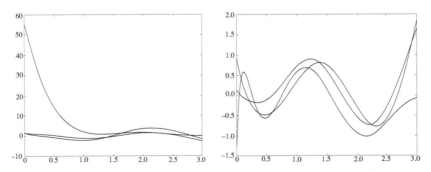

Figure 4.3.4 $\bar{y}_\epsilon(t)$ and $a_\epsilon(t)$ for example 4.3.1: three simultaneous hyperplane tests.

For simplicity of discussion we shall take the system matrices as constants. However, the discussion and algorithms go through with only the obvious changes if the system matrices are continuous functions of t.

Detection is to be carried out over an interval $[0, T]$. The initial condition $x_i(0)$ is unknown and treated as another disturbance. We suppose that the disturbances $\nu_i, x_i(0)$ satisfy

$$S(\nu_i, x_i(0)) = x_i(0)^T \hat{P}_i x_i(0) + \int_0^T |\nu_i(t)|^2 dt \leq 1 \qquad (4.4.1)$$

and that there are m models $i = 0, \ldots, m-1$. As will become clear later, the approach of this section does not apply when $\hat{P}_i$ is singular. Accordingly we assume that $\hat{P}_i > 0$. Note that (4.4.1) implies that

$$|\hat{P}^{-1/2} x_i(0)| < 1, \quad \|\nu\|^2 = \int_0^T |\nu_i|^2 dt < 1.$$

Again let $\mathcal{A}_i(v)$ be the set of outputs for (4.2.1) given that (4.4.1) holds. We have given algorithms for computing a proper v on $[0, T]$. In addition for each pair of models i, j for which a decision is desired, algorithms have been given for computing the functions $a_{i,j}(t), \bar{y}_{i,j}(t)$ and the constants $\bar{\epsilon}_{i,j} \geq 0, \underline{\epsilon}_{i,j} \leq 0$, so that if we define the auxiliary signal

$$\psi_{i,j}(y) = \int_0^T (y(s) - \bar{y}_{i,j}(s))^T a_{i,j}(s) ds, \qquad (4.4.2)$$

then

$$\text{if } y \in \mathcal{A}_i(v) \cup \mathcal{A}_j(v), \text{ then either } \begin{cases} \psi_{i,j}(y) \geq \bar{\epsilon}_{i,j} \text{ and } y \in \mathcal{A}_i(v) \\ \qquad\qquad \text{or} \\ \psi_{i,j}(y) \leq \underline{\epsilon}_{i,j} \text{ and } y \in \mathcal{A}_j(v). \end{cases} \qquad (4.4.3)$$

Everything but the integral in (4.4.2) is precomputed and the integral can be computed as y is received. Thus the test can be carried out in real time. If there are only two models and a minimum proper auxiliary signal v is used, then $\bar{\epsilon}_{1,2} = \underline{\epsilon}_{1,2} = 0$. However, the test is more robust, and decisions can often be made more quickly, if a proper v of larger norm is used so that

$$\bar{\epsilon}_{1,2} > 0 > \underline{\epsilon}_{1,2}. \qquad (4.4.4)$$

The strict inequality (4.4.4) usually holds for most of the tests for more than two models even if v is a minimum proper auxiliary signal.

When there are more than two models the decision making is slightly more complex since the output y might be coming from a model different from the two models used in the i, j comparison. If there are m models, then the decision becomes

$$\text{if } y \in \cup_{p=0}^{m-1} \mathcal{A}_p(v), \text{ then } \begin{cases} \psi_{i,j}(y) < \bar{\epsilon}_{i,j} \Rightarrow y \notin \mathcal{A}_i(v) \\ \psi_{i,j}(y) > \underline{\epsilon}_{i,j} \Rightarrow y \notin \mathcal{A}_j(v). \end{cases} \qquad (4.4.5)$$

The test (4.4.5) is illustrated in figure 4.4.1 for the case of four models and a comparison based on models $i = 1, j = 2$.

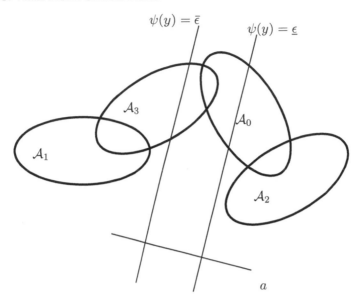

$$\psi(y) = \bar{\epsilon}$$

$$\psi(y) = \underline{\epsilon}$$

Figure 4.4.1 Four output sets and the $1, 2$ separating hyperplanes.

Some notation and terminology are needed. The L^2 norm of a vector function restricted to $[\tau, T]$ is

$$\|z\|_{[\tau,T]} = \left(\int_{\tau}^{T} |z(t)|^2 dt \right)^{1/2} \tag{4.4.6}$$

so that $\|z\| = \|z\|_{[0,T]}$. A matrix-valued function $H(t)$ as an operator on L^2 has a norm that is the supremum over t of the matrix norm of $H(t)$. We denote this $\|H\|$. We define the operator $\mathcal{L}_i$ by

$$\mathcal{L}_i(f) = \int_{0}^{t} e^{A_i(t-s)} f(s) ds$$

so that $\mathcal{L}_i(f)$ is the unique solution of $\dot{x} = A_i x + f$, $x(0) = 0$. Given a normal $a_{i,j}$, we may divide it by $\|a_{i,j}\|$ and it is still a normal, so we may assume that $\|a_{i,j}\| = 1$. Let

$$\Psi_{i,j}(y, \tau) = \int_{0}^{\tau} a_{i,j}(s)^T (y(s) - \bar{y}_{i,j}(s)) ds.$$

We assume that the integral defining $\psi_{i,j}$ is computed as y is measured. Thus $\Psi_{i,j}(y, \tau)$ is the value of $\psi_{i,j}(y)$ computed up to time τ. Suppose that $y \in \mathcal{A}_p(v)$ but we do not know what p is. Then

$$y = C_p \mathcal{L}_p B_p v + C_p e^{A_p t} x_p(0) + (C_p \mathcal{L}_p M_p + N_p) \nu_p$$

so that

$$\psi_{ij}(y) = \Psi_{i,j}(y, \tau) + \Delta_{i,j}(y, \tau),$$

where

$$\Delta_{i,j}(y,\tau) = \int_\tau^T a_{ij}^T \left(C_p \mathcal{L}_p B_p v + C_p e^{A_p s} x_p(0) + (C_p \mathcal{L}_p M_p + N_p) \nu_p - \bar{y}_{i,j} \right) ds. \tag{4.4.7}$$

In estimating $\Delta_{i,j}(t)$ a smaller value is obtained by grouping as many terms together as possible. Rewrite (4.4.7) as

$$\Delta_{i,j}(\tau) = \int_\tau^T a_{ij}^T (C_p \mathcal{L}_p B_p v - \bar{y}_{i,j}) ds + \int_\tau^T a_{ij}^T (C_p \mathcal{L}_p M_p + N_p) \nu_p ds$$
$$+ \int_\tau^T a_{ij}^T C_p e^{A_p s} x_p(0) ds. \tag{4.4.8}$$

From the Cauchy-Schwartz inequality we have that for any L^2 functions g, h

$$\left| \int_\tau^T g(s)^T h(s) ds \right| \le \left(\int_\tau^T |g(s)|^2 ds \right)^{1/2} \left(\int_\tau^T |h(s)|^2 ds \right)^{1/2} = \|g\|_{[\tau,T]} \|h\|_{[\tau,T]}.$$

The following quantities and scalar functions can all be computed off-line as will be shown later:

$$\delta_{i,j}(\tau) = \left(\int_\tau^T \|a_{i,j}(s)\|^2 ds \right)^{1/2} = \|a_{i,j}\|_{[\tau,T]}, \tag{4.4.9}$$

$$K_i(\tau) = \|C_i \mathcal{L}_i M_i + N_i\|_{[\tau,T]}, \tag{4.4.10}$$

$$\gamma_{i,j,k}(\tau) = \left| \int_\tau^T a_{i,j}(s)^T \left(C_k \mathcal{L}_k B_k v(s) - \bar{y}_{i,j}(s) \right) ds \right|, \tag{4.4.11}$$

$$\theta_k(\tau) = \left(\int_\tau^T \|C_k e^{A_k s} \hat{P}_k^{-1/2}\|^2 ds \right)^{1/2}. \tag{4.4.12}$$

Notice that $\delta_{i,j}, \theta_k$ are monotonically decreasing functions which are zero at T. $\gamma_{i,j,k}$ also goes to zero as t goes to T, and $\lim_{t \to T^-} K_i(t) = |N_i|$. In an actual implementation, things may be simplified by replacing most terms by larger but simpler functions.

Since $\mathcal{S}(x_p(0), \nu_p) \le 1$ we get

$$|\Delta_{i,j}(y,\tau)| \le \gamma_{i,j,p}(\tau) + (K_p(\tau) + \theta_p(\tau)) \delta_{i,j}(\tau). \tag{4.4.13}$$

Notice that the previous calculation uses $\hat{P}_i > 0$. If $\hat{P}_i$ is not invertible, then we cannot get a bound on the term $e^{A_p t} x(0)$ in (4.4.8) prior to $t = T$.

From (4.4.13) we get the following theorem on early decision.

Theorem 4.4.1 *Suppose that v is a proper m-model auxiliary signal. Suppose that $y \in \cup_{p=0}^{m-1} \mathcal{A}_p(v)$. Then*

1. *If for some $0 < t < T$ we have that*

$$\Psi_{i,j}(y,t) > \epsilon_{ij} + \max_p \left[\gamma_{i,j,p}(t) + (K_p(t) + \theta_p(t)) \delta_{i,j}(t) \right], \text{ then } y \notin \mathcal{A}_j(v);$$

$$\tag{4.4.14}$$

2. *If for some $0 < t < T$ we have that*

$$\Psi_{i,j}(y,t) < \bar{\epsilon}_{ij} - \max_p \left[\gamma_{i,j,p}(t) + (K_p(t) + \theta_p(t)) \, \delta_{i,j}(t) \right], \text{ then } y \notin \mathcal{A}_i(v).$$
(4.4.15)

All the signals used in the test can be precomputed and a piecewise polynomial upper bound suffices. Thus this modification of the original hyperplane test uses little extra memory and essentially no extra computation that is not off-line.

It is also useful to note that if we have more than two models and one of the estimates (4.4.13) is larger than the others and that test condition is met, then we need no longer use that test as part of the remaining tests. That is, we can modify the tests as follows.

During a given test period let $\Sigma(t)$ be the indices of the possible models that might still be correct at time t. Thus as soon as a test reveals that, say, $y \notin \mathcal{A}_k(v)$, the integer k is deleted from Σ. Then $\Sigma(0) = \{0, 1, \dots, m-1\}$ and $\Sigma(T)$ is a singleton which depends on the model giving y. Then (4.4.14) and (4.4.15) become

$$\text{if } \Psi_{i,j}(y,t) > \underline{\epsilon}_{ij} + \max_{p \in \Sigma(t)} \left[\gamma_{i,j,p}(t) + (K_p(t) + \theta_p(t)) \, \delta_{i,j}(t) \right], \text{ then } y \notin \mathcal{A}_j(v),$$
(4.4.16)

$$\text{if } \Psi_{i,j}(y,t) < \bar{\epsilon}_{ij} - \max_{p \in \Sigma(t)} \left[\gamma_{i,j,p}(t) + (K_p(t) + \theta_p(t)) \, \delta_{i,j}(t) \right], \text{ then } y \notin \mathcal{A}_i(v).$$
(4.4.17)

4.4.1 Computation of Needed Terms

We now turn to addressing the computation of the needed terms. We choose to compute the needed quantities in a way that allows the use of standard software packages. Since functions are needed we will make use of differential equation integration software which has the capability for continuous output. That is, instead of just returning values of the solution of the differential equation at isolated points, the software returns a function for the solution which is piecewise polynomial. We assume then that one already has the functions $a_{i,j}, \bar{y}, v$. Formulas are given for the case where the system matrices are constant. The formulas are easily modified for the time-varying case.

$\delta_{i,j}$ is computed by solving the scalar differential equation

$$\dot{z}(t) = -a_{i,j}^T(t) a_{i,j}(t), \quad z(T) = 0,$$
(4.4.18a)

$$\delta_{i,j}(t) = z^{1/2}(t).$$
(4.4.18b)

$\gamma_{i,j,k}$ is computed from the boundary value problem

$$\dot{x}(t) = A_k x(t) + B_k v(t), \quad x(0) = 0,$$
(4.4.19a)

$$\dot{z}(t) = -a_{i,j}^T(t) \left(C_k x(t) - \bar{y}_{i,j}(t) \right), \quad z(T) = 0,$$
(4.4.19b)

$$\gamma_{i,j,k}(t) = |z(t)|.$$
(4.4.19c)

θ_k is easily computed for moderately sized problems by

$$\dot{z}(t) = -|C_k e^{At} P^{-1/2}|^2, \quad z(T) = 0,$$
(4.4.20a)

$$\theta_k(t) = z^{1/2}(t).$$
(4.4.20b)

This leaves only the K_i term in (4.4.10). In the general case the $M_i\nu$ and $N_i\nu$ are separate noise terms so not much is lost by estimating terms involving N_i, M_i separately.

Consider a particular $t \in [0,T]$. Then, suppressing the subscripts, we have

$$|C\mathcal{L}M(\nu)(t)| \le \left| \int_0^t Ce^{A(t-s)}M\nu(s)\,ds \right| \quad \text{for } t \in [0,T] \tag{4.4.21}$$

$$\le \int_0^t |Ce^{A(t-s)}M||\nu(s)|\,ds \tag{4.4.22}$$

$$\le \left(\int_0^t |Ce^{A(t-s)}M|^2\,ds \right)^{1/2} \left(\int_0^t |\nu(s)|^2\,ds \right)^{1/2} \tag{4.4.23}$$

$$\le \left(\int_0^t |Ce^{At}M|^2\,ds \right)^{1/2} \tag{4.4.24}$$

since $s \le t$. But

$$\|C\mathcal{L}M\|_{[t,T]} \le \left(\int_t^T \left(\left(\int_0^s |Ce^{Ar}M|^2 dr \right)^{1/2} \right)^2 ds \right)^{1/2} \tag{4.4.25}$$

$$= \left(\int_t^T \left(\int_0^s |Ce^{Ar}M|^2 dr \right) ds \right)^{1/2}. \tag{4.4.26}$$

Thus K_i may be computed as follows. Let $\bar{M}_i(t)$ be such that
$$|C_i e^{A_i t}M_i| \le \bar{M}_i(t) \quad \text{for all } t \in [0,T].$$

Then

$$\dot{z} = \bar{M}_i^2, \quad z(0) = 0, \tag{4.4.27a}$$
$$\dot{q} = -z, \quad q(T) = 0, \tag{4.4.27b}$$
$$K_i(t) = |N_i| + q_i^{1/2}(t). \tag{4.4.27c}$$

4.4.2 Early Detection E xample

While the algorithms developed earlier can easily handle much more complex problems, we give here one simple computational example that illustrates several key points about early detection. Consider (4.2.34) of example 4.2.3. We take $\hat{P}_i = I$ and $T = 20$. Quicker detection is possible if one uses $\rho\nu$ with $\rho > 1$. We shall use $\rho = 1$.

Since we will use the minimum proper ν and there are only two models, we have $\underline{\epsilon}_{01} = 0$ and $\bar{\epsilon}_{01} = 0$. Let

$$\widehat{\delta}(t) = \max_{p=0,1} \left[\gamma_{0,1,p}(t) + (K_p(t) + \theta_p(t))\,\delta_{0,1}(t) \right].$$

Then tests (4.4.14) and (4.4.15) of theorem 4.4.1 become

1. If for some $0 < t < T$ we have that
$$\Psi(y,t) > \widehat{\delta}(t), \text{ then } y \notin \mathcal{A}_1(v); \tag{4.4.28}$$

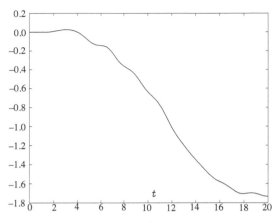

Figure 4.4.2 $\Psi_{0,1}(y_0, t)$ for one sample run.

2. If for some $0 < t < T$ we have that

$$\Psi(y, t) < -\widehat{\delta}(t), \text{ then } y \notin \mathcal{A}_0(v). \tag{4.4.29}$$

To evaluate the new test we performed the following experiment . We took $T = 20$ and generated 100 random noise signals ν_i and initial conditions. Then for each noise signal and initial condition we ran simulations assuming first that model 0 was correct and then that model 1 was correct. There is some question about what exactly a random $x_i(0)$ and a random ν_i are. We proceeded to generate them as follows. We generated uniformly on $[-1, 1]$ the first 50 Fourier sine coefficients and the initial conditions. These were renormed to be norm 1 in R^n and L^2. We then randomly generated three numbers uniformly distributed on $[0, 1]$ whose norm was less than 1 to multiply the initial conditions and the two noises.

A typical value of $\Psi(y_0, t)$ for one sample run is shown in figure 4.4.2. Note that Ψ undergoes a change in sign before finally becoming definite enough to permit early detection.

The two graphs in figure 4.4.3 show the results of these 200 trials. Each graph gives the 100 different $\Psi(y_0, t)$ curves and the graph of $\pm \hat{\delta}(t)$. The $\pm \hat{\delta}$ are the two curves that converge to zero at $T = 20$. Detection occurs when the Ψ graph crosses the $\hat{\delta}$ graph.

For model 0 detection time ranged from 14.444 to 15.9722 with an average of 15.028. For model 1 detection time ranged from 14.0278 to 15.278 with an average of 14.651. The actual distributions of detection times are shown in figure 4.4.4. Thus, for example, figure 4.4.4 shows that, when model 0 was correct, over 60 of the 100 trials resulted in detection by $t = 15.0$.

The jumps in figure 4.4.4 are due to the step size of the simulation. The numerical method is fourth order and was able to integrate with steps around 0.1. A smoother curve would result if a root finding option was used to find out when $\hat{\delta}(t)$ was crossed. However, the observations below would be unchanged. Notice the clustering of the detection times at around $0.75T$. This is due to two factors. First, since $\hat{\delta}$ is an

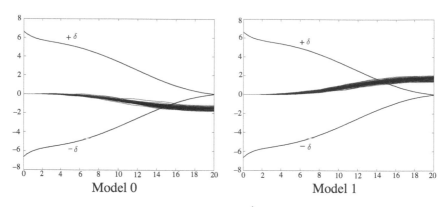

Figure 4.4.3 $\Psi(y_0, t)$ and $\pm\hat{\delta}(t)$ for 200 trials.

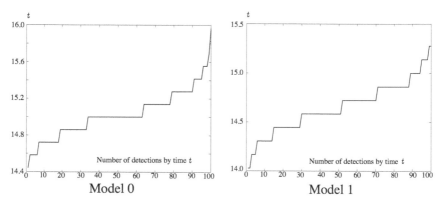

Figure 4.4.4 Distribution of detection times for 200 trials.

estimate it is very difficult to make detection in the early part of the time interval. On the other hand, the worst case noise is also highly unlikely. Since it is also the solution of an optimization problem, the worst case noise takes a special form, and randomly generated noises of varying norm are usually not close to the worst case in either norm or form.

4.5 OTHER EXTENSIONS

There are a number of other problems that are conceptually straightforward to formulate using the methods of this chapter but are more difficult, or impossible, to solve using the solution techniques of previous chapters. We will briefly mention some of these extensions.

4.5.1 Other Inputs

Often in control problems there are open loop controls u, such as reference signals, which are known ahead of time. Suppose that we now want to apply an auxiliary signal v to provide model identification. Suppose also that our auxiliary signal v is applied through the same channels as u. Then the models take the form

$$\dot{x}_i = A_i x_i + B_i(u + v) + M_i \nu_i, \tag{4.5.1a}$$

$$y = C_i x_i + N_i \nu_i. \tag{4.5.1b}$$

The u of (4.5.1) should not be confused with the on-line measured u of chapter 3. In chapter 3 the on-line measured u was a signal that only became known as the test progressed and thus could not be used in the design of v. Here u is a fixed signal that is known ahead of time and can be used to find v.

The optimization problem of finding a minimum energy proper auxiliary signal using optimization formulation 4 (4.2.25) is now

$$\min_v \delta^2(v) \text{ such that } \phi_\beta(u + v) \geq 1 \text{ for some } \beta \in [0, 1]. \tag{4.5.2}$$

Note that $v = 0$ is possible if u is already proper. Example 4.6.1 is of the form (4.5.2).

If the reference signal is known ahead of time and is not coming from the same channels as the auxiliary signal, then we have

$$\dot{x}_i = A_i x_i + B_i v + E_i u + M_i \nu_i, \tag{4.5.3a}$$

$$y = C_i x_i + N_i \nu_i, \tag{4.5.3b}$$

and u is considered as known. One could go back and derive this from first principles but it is easier to proceed as follows. We combine u, v into a new input vector $w = [w_1, w_2] = [u, v]$. For this problem we may construct $\phi_\beta(w)$ as before. The optimization problem then becomes

$$\min_w \delta(w_2)^2 \text{ such that } \phi_\beta(w) \geq 1 \text{ for some } \beta \in [0, 1] \text{ and } w_1 = u. \tag{4.5.4}$$

That is, the problem is similar in appearance to the preceding ones except that the cost depends only on the w_2 part of the test signal w, and we have an additional constraint on the w_1 part of w.

4.5.2 Nonlinearities

The optimization approach can, in principle, handle a number of different types of nonlinearities but, for fundamental reasons, only nonlinearities entering the system in a special way can be considered. The actual computational success will depend, as always, on the particular nonlinearity.

Nonlinear in the Control

Suppose that the control enters the equations in a nonlinear way so that the models are

$$\dot{x}_i = A_i x_i + B_i g(v) + M_i \nu_i, \tag{4.5.5a}$$

$$y = C_i x_i + N_i \nu_i. \tag{4.5.5b}$$

For example, if v is a steering angle it may enter the equations as $\cos v$. The optimization problem using formulation 4 is now

$$\min_v \delta(v)^2 \text{ such that } \phi_\beta(g(v)) \geq 1 \text{ for some } \beta \in [0,1]. \qquad (4.5.6)$$

This gives us a nonlinear boundary value problem. The nature of the nonlinearity will impact, of course, on the solution of the optimization problem. If the function g is not one to one, then it is possible that there might be multiple local minima and saddle points. This may slow convergence unless good initial guesses are used.

More generally than (4.5.5), the optimization formulations 2 and 4 are still correct if we have

$$\dot{x}_i = A_i(v)x_i + B_i(v) + M_i(v)\nu_i, \qquad (4.5.7a)$$
$$y = C_i(v)x_i + N_i(v)\nu_i \qquad (4.5.7b)$$

instead of (4.5.5). The reason that this works is that for a fixed v the problem of computing $\phi_\beta(v)$ is still a constrained linear quadratic regulator problem. In particular, $A_i(v)$ is still convex for each v.

However, when the optimization software goes to optimize $\delta(v)^2$ over v, the problem becomes fully nonlinear, and one is faced with all the problems typical of nonlinear control problems including the possibility of multiple local minima and finding good initial guesses. On the other hand, any of the solutions found will be proper. Often what is important in applications is the usefulness of the signal and not whether it is actually optimal. Note the comments later in subsection 4.5.4 on implementation of auxiliary signals.

Small Nonlinearities

As is the case with many robustness algorithms, we can handle small norm bounded nonlinearities by increasing the noise bound. For example, suppose that the models and noise bounds are of the form

$$\dot{x}_i = A_i x_i + g_i(x_i, t) + B_i v + M_i \nu_i, \qquad (4.5.8a)$$
$$y = C_i x_i + N_i \nu_i, \qquad (4.5.8b)$$
$$\|\nu_i\|^2 < 1. \qquad (4.5.8c)$$

If $|g_i(x,t)| \leq \delta$ for all (t,x), then (4.5.8) can be attacked by considering

$$\dot{x}_i = A_i x_i + B_i v + \bar{M}_i \bar{\nu}_i, \qquad (4.5.9a)$$
$$y = C_i x_i + \bar{N}_i \bar{\nu}, \qquad (4.5.9b)$$
$$\|\bar{\nu}_i\|^2 < 1 + \delta^2 T, \qquad (4.5.9c)$$

which can be rescaled to be in the form of (4.2.1). Here $\bar{\nu} = (\hat{\nu}_i, \nu_i)$, $\bar{M}_i = \begin{pmatrix} I & M_i \end{pmatrix}$, and $\bar{N}_i = \begin{pmatrix} 0 & N_i \end{pmatrix}$. Approaching the problem this way will produce a conservative answer.

4.5.3 Additional Bounds on v

In some applications one wants to put more restrictions on the auxiliary signal than just minimizing its norm. For example, for design purposes, one might want a proper signal of smallest energy which also satisfies a pointwise bound, say $L_1 \leq v(t) \leq L_2$, for safety or other considerations. Also, in some applications it only makes sense for a control to be nonnegative. These kinds of mixed signal constraints are difficult to handle using the approach of chapter 3 but can easily be accommodated within the optimization approach of this chapter. For example, if the constraint is $L_1 \leq v(t) \leq L_2$, the two-model formulation of this problem will be

$$\min \delta(v)^2, \tag{4.5.10a}$$

$$\phi_\beta(v) \geq 1, \tag{4.5.10b}$$

$$L_1 \leq v(t) \leq L_2 \text{ for all } t \in [0, T]. \tag{4.5.10c}$$

Of course, there may not be a proper signal that meets the additional constraint (4.5.10c).

Example 4.5.1 *Consider the two models in (4.2.3) with no weight on the initial state and $U = 0, W = 0$. If $T = 1$ we get the tallest auxiliary signal in figure 4.2.8. However, suppose that we have the additional pointwise bound on v,*

$$|v(t)| \leq 65. \tag{4.5.11}$$

Then the minimum energy proper auxiliary signal found by solving (4.5.10) is given by the graph in figure 4.5.1.

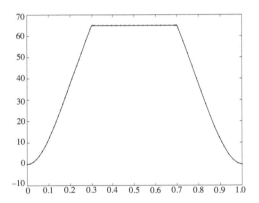

Figure 4.5.1 Minimum proper v with pointwise bound (4.5.11).

An interesting variant of this problem is to suppose that a pointwise bound such as (4.5.10c) is specified and then to try and find the shortest test interval for which there exists a proper auxiliary signal that meets the bound. This leads to the optimization problem

$$\min T, \tag{4.5.12a}$$

$$\phi_\beta(v) \geq 1, \tag{4.5.12b}$$

$$L_1 \leq v(t) \leq L_2 \text{ for all } t \in [0, T]. \tag{4.5.12c}$$

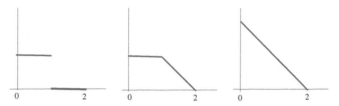

Figure 4.5.2 The auxiliary signal shapes (4.5.13), (4.5.14), and (4.5.15), respectively.

4.5.4 Implementing an Auxiliary Signal

Implementing a complex auxiliary signal may not be desirable for a number of reasons. It is often easier to apply signals which are either piecewise constant or piecewise linear. One might look for proper auxiliary signals of the form cw where w is some simple to implement auxiliary signal and c is a scalar to be determined. Note that, for this type of problem, the optimization software has a much simpler task since there is only an unknown parameter c rather than an unknown signal v. The infinite-dimensional optimization problem in v has become a finite-dimensional optimization problem in c. The optimization problem is

$$\min |c| \text{ such that } \phi_\beta(cw) \geq 1.$$

Example 4.5.2 *To illustrate, suppose that one wishes to implement a simpler auxiliary signal than that shown later in figure 4.6.2. The graphs in figure 4.6.2 motivate considering the following three simple auxiliary signals of shape w:*

$$w_1(t) = \begin{cases} 1 & \text{if} \quad 0 \leq t < 1 \\ 0 & \text{if} \quad 1 < t < 2, \end{cases} \tag{4.5.13}$$

$$w_2(t) = \begin{cases} 1 & \text{if} \quad 0 \leq t \leq 1 \\ 2 - t & \text{if} \quad 1 \leq t \leq 2, \end{cases} \tag{4.5.14}$$

and

$$w_3(t) = 2 - t. \tag{4.5.15}$$

These three signal shapes are shown in figure 4.5.2.

Suppose that we wish to minimize $\|cw\|$. Computing the minimum c we get that $\|cw_1\| = 16.47$, $\|cw_2\| = 15.42$, and $\|cw_3\| = 15.57$. We see that if, instead of using a conventional step input like w_1, we use a shape more like the optimal shape, we can then use smaller auxiliary signals.

If the auxiliary signal is r dimensional, then we can consider $[c_1 v_1, \ldots, c_r v_r]$ and there are r parameters to minimize over.

Another way to reduce the problem to a problem with a finite number of parameters is to assume v is piecewise constant. Then the constant values, and possibly also the jump times, are the parameters. Numerous other options are possible.

4.6 SYSTEMS WITH DELAYS

Many systems have models for which the state space is infinite dimensional. These include models with partial differential equations and those with delays. There are two general approaches to auxiliary signal design for such problems. One is to return to first principles and try to develop a theory and algorithms based on this theory. The second is to approximate the infinite-dimensional system by a finite-dimensional one and then try to apply the techniques that were previously developed in this book to the finite-dimensional problem. In this section we give an example of the second approach by considering delay systems.

Delays occur in many applications so it is natural to want to perform failure detection on models with delays. We first focus on a special class of delay problems which can be reformulated so that they can be solved using modifications of the existing theory. This case arises in some applications and serves as a truth model for the approximation approaches discussed later. The reformulation poses some new technical issues in terms of prior information. It also requires considering different boundary conditions. The inclusion of general variable delays or multiple delays in the models being identified results in a problem which is intrinsically infinite dimensional and requires additional theoretical and numerical results. These can be handled by approximating the delay system by an ordinary differential equation [45], much as is done with the method of lines for partial differential equations. A detailed discussion of this is beyond the scope of this book; however, there is some introductory discussion in section 4.6.3. All the approaches that we discuss require some care in formulation in order to be solved correctly. We shall consider only the case of additive noise. The methods given in this section adapt easily to the model uncertainty case. However, the supporting analysis only applies to model uncertainties that are small enough to give bounded sets of admissible noises [19].

4.6.1 Method of Steps

We assume that we have m possible models of the form

$$\dot{x}_i(t) = A_i x_i(t) + G_i x_i(t - h) + B_i v(t) + M_i \nu_i(t), \qquad (4.6.1a)$$

$$y(t) = C_i x_i(t) + N_i \nu_i(t), \qquad (4.6.1b)$$

$$x_i(t) = \phi_i(t), \quad -h \le t < 0, \qquad (4.6.1c)$$

for $i = 0, \dots, m - 1$, where v is an auxiliary signal to be determined and y is the observed output. For simplicity of discussion we take the system matrices as constants; however, the discussion goes through with only the obvious changes if the system matrices are continuous functions of t. We do not know the prior history ϕ_i and thus it is a type of disturbance or noise as are the ν_i and the initial condition $x_i(0)$. Note that the delay is the same for all models. We shall discuss the case when the change in delay may be part of the failure later in this section.

Detection is to be carried out over a finite interval $[0, T]$. We suppose that the disturbances $\phi_i, \nu_i, x_i(0)$ satisfy

$$\mathcal{S}_i(\phi_i, \nu_i, x_i(0)) = x_i(0)^T Q_i x_i(0) + \int_{-h}^{0} |\phi_i|^2 dt + \int_{0}^{T} |\nu_i|^2 dt \le 1 \qquad (4.6.2)$$

where $Q_i \geq 0$. This is a direct generalization of (4.2.2). Let $\mathcal{A}_i(v)$ be the set of outputs y for (4.6.1) given that (4.6.2) holds.

We focus on determining the auxiliary signal. The modifications necessary for the hyperplane test are then clear. We also focus on the case with one delay and two models since that suffices to discuss the points we wish to make.

If $T \leq h$, then we can alter (4.6.2) to

$$S(\phi_i, \nu_i, x_i(0)) = x_i(0)^T Q_i x_i(0) + \int_{-h}^{T-h} |\phi_i|^2 dt + \int_0^T |\nu_i|^2 dt \leq 1 \quad (4.6.3)$$

and treat ϕ_i on $[-h, T-h]$ as part of the initial disturbance. That is, $\begin{pmatrix} \phi \\ \nu_i \end{pmatrix}$ is the new ν_i and $\begin{pmatrix} G_i & M_i \end{pmatrix}$ is the new M_i. This is now the case considered earlier in this chapter and no special consideration is needed. Thus we assume that $T > h$. For convenience, we assume that $T = \bar{\kappa} h$ for an integer $\bar{\kappa} \geq 1$. This is not restrictive for h small relative to T. For h large relative to the desired T this becomes somewhat restrictive. This problem can be handled but it requires some modifications and will not be discussed here.

Let $\kappa = \bar{\kappa} - 1$ and assume that (4.6.1) holds for $i = 0, 1$ to give us the combined system

$$\dot{x}_0(t) = A_0 x_0(t) + G_0 x_0(t-h) + B_0 v(t) + M_0 \nu_0(t), \quad (4.6.4a)$$
$$y(t) = C_0 x_0(t) + N_0 \nu_0(t), \quad (4.6.4b)$$
$$\dot{x}_1(t) = A_1 x_1(t) + G_1 x_1(t-h) + B_1 v(t) + M_1 \nu_1(t), \quad (4.6.4c)$$
$$y(t) = C_1 x_1(t) + N_1 \nu_1(t). \quad (4.6.4d)$$

Elimination of y gives the delay descriptor system

$$\dot{x}_0(t) = A_0 x_0(t) + G_0 x_0(t-h) + B_0 v(t) + M_0 \nu_0(t), \quad (4.6.5a)$$
$$\dot{x}_1(t) = A_1 x_1(t) + G_1 x_1(t-h) + B_1 v(t) + M_1 \nu_1(t), \quad (4.6.5b)$$
$$0 = C_0 x_0(t) - C_1 x_1(t) + N_0 \nu_0(t) - N_1 \nu_1(t). \quad (4.6.5c)$$

We now convert (4.6.5) to an ordinary differential equation without delays. Let $z_{i,j}$ on $[0, h]$ be given by $z_{i,j}(t) = x_i(t + jh)$ for $i = 0, 1$ and $\kappa \geq j \geq 0$. Then $\mu_{i,j}$ is defined similarly from ν_i. For $0 \leq t \leq h$, let $\gamma_i(t) = \phi_i(t-h)$. Let $v_j(t) = v(t+jh)$. Then we can rewrite the system (4.6.5) as one without delay on $[0, h]$ as shown in (4.6.6):

$$\dot{z}_{0,0} = A_0 z_{0,0} + G_0 \gamma_0 + B_0 v_0 + M0\mu_{0,0}, \quad (4.6.6a)$$
$$\dot{z}_{0,1} = A_0 z_{0,1} + G_0 z_{0,0} + B_0 v_1 + M_0 \mu_{0,1}, \quad (4.6.6b)$$
$$\vdots \quad\quad\quad\quad\quad\quad\quad\quad\quad\quad\quad (4.6.6c)$$

$$\dot{z}_{0,\kappa} = A_0 z_{0,\kappa} + G_0 z_{0,\kappa-1} + B_0 v_\kappa + M_0 \mu_{0,\kappa}, \qquad (4.6.6\text{d})$$

$$\dot{z}_{1,0} = A_1 z_{1,0} + G_1 \gamma_1 + B_1 v_0 + M_1 \mu_{1,0}, \qquad (4.6.6\text{e})$$

$$\dot{z}_{1,1} = A_1 z_{1,1} + G_1 z_{1,0} + B_1 v_0 + M_1 \mu_{1,1}, \qquad (4.6.6\text{f})$$

$$\vdots \qquad (4.6.6\text{g})$$

$$\dot{z}_{1,\kappa} = A_1 z_{1,\kappa} + G_1 z_{1,\kappa-1} + B_1 v_\kappa + M_1 \mu_{1,\kappa}, \qquad (4.6.6\text{h})$$

$$0 = C_0 z_{0,0} - C_1 z_{1,0} + N_0 \mu_{0,0} - N_1 \mu_{1,0}, \qquad (4.6.6\text{i})$$

$$\vdots \qquad (4.6.6\text{j})$$

$$0 = C_0 z_{0,\kappa} - C_1 z_{1,\kappa} + N_0 \mu_{0,\kappa} - N_1 \mu_{1,\kappa} \qquad (4.6.6\text{k})$$

along with the initial and boundary conditions

$$z_{i,0}(0) = x_i(0) \quad i = 0, 1, \qquad (4.6.7\text{a})$$

$$z_{i,j}(0) = z_{i,j-1}(h), \quad i = 0, 1, \quad j = 1, \dots, \kappa. \qquad (4.6.7\text{b})$$

In the previous sections when working without delays there were two options on how to handle the initial conditions. Depending on the problem one could take the weight either positive semidefinite or semidefinite including zero. It was a design choice. Here the situation is somewhat different. To simplify the discussion the only semidefinite case we will consider is zero weight.

If G_i is invertible, then for bounded noise and any fixed v, we see from (4.6.6a), (4.6.6e) that we can get any smooth $z_{i,0}$. But then from (4.6.6b) we can get any $z_{i,1}$ since $z_{i,0}$ is arbitrary. Continuing in this manner we see that if G_i is invertible, and there is no information on ϕ_i, then model identification is not possible. Thus ϕ_i has to be included in (4.6.2) with a positive definite weight.

What about $x_i(0)$? There are now three possibilities. One option is that $x_i(0)$ is free and $Q_i = 0$ in (4.6.2). A second option is that $x_i(0)$ is viewed as another perturbation treated like ν_i and we have $Q_i > 0$ in (4.6.2). These are analogous to the options we had before. However, there is now a third option. In this option we assume that the system has been operating prior to the test. Thus while we do not know ϕ_i we do have the additional continuity boundary condition

$$z_{i,0}(0) = \phi_i(0). \qquad (4.6.8)$$

The boundary condition (4.6.8) is somewhat unusual. In the algorithms that follow we will minimize the noise measure subject to some constraints. In so doing the noises will be treated as controls. But then (4.6.8) is a boundary condition that links the state to the control. This is not a typical restriction on the control and can result in numerical ill-conditioning. We will not discuss this option further.

For $q = z_0, z_1, v$, let $\bar{q}$, be the vector with components $q_0, q_1, \dots, q_\kappa$ and $\bar{q}_i$ the vector with components $q_{i,0}, \dots, q_{i,\kappa}$). The vector $\bar{\mu}_i$ is defined the same way except that we add γ_i to the first component. Then we get that (4.6.6) is the system (4.6.9) on $[0, h]$:

$$\dot{\bar{z}}_0 = \bar{A}_0 \bar{z}_0 + \bar{B}_0 \bar{v} + \bar{M}_0 \bar{\mu}_0, \qquad (4.6.9\text{a})$$

$$\dot{\bar{z}}_1 = \bar{A}_1 \bar{z}_1 + \bar{B}_1 \bar{v} + \bar{M}_1 \bar{\mu}_1, \qquad (4.6.9\text{b})$$

$$0 = \bar{C}_0 \bar{z}_0 - \bar{C}_1 \bar{z}_1 + \bar{N}_0 \bar{\mu}_0 - \bar{N}_1 \bar{\mu}_1, \qquad (4.6.9\text{c})$$

where

$$\bar{Q}_i = \text{diag}\{Q_i, \dots, Q_i\}, \quad \bar{A}_i = \begin{pmatrix} A_i & 0 & 0 & 0 \\ G_i & A_i & 0 & \ddots \\ 0 & \ddots & \ddots & \ddots \\ 0 & 0 & G_i & A_i \end{pmatrix}, \tag{4.6.10a}$$

$$\bar{z}_i = \begin{pmatrix} z_{i,0} \\ \vdots \\ z_{i,\kappa} \end{pmatrix}, \quad \bar{\mu}_i = \begin{pmatrix} \gamma_i \\ \mu_{i,0} \\ \vdots \\ \mu_{i,\kappa} \end{pmatrix}, \quad \bar{v} = \begin{pmatrix} v_0 \\ \vdots \\ v_\kappa \end{pmatrix}, \tag{4.6.10b}$$

$$\bar{M}_i = \begin{pmatrix} G_i & M_i & 0 & \cdot \\ 0 & 0 & \ddots & \cdot \\ 0 & 0 & 0 & M_i \end{pmatrix}, \quad \bar{N}_i = \begin{pmatrix} 0 & N_i & 0 & \cdot \\ 0 & 0 & \ddots & \cdot \\ 0 & 0 & 0 & N_i \end{pmatrix}, \tag{4.6.10c}$$

$$\bar{B}_i = \text{diag}\{B_i, B_i, \dots, B_i\}, \quad \bar{C}_i = \text{diag}\{C_i, C_i, \dots, C_i\}. \tag{4.6.10d}$$

Using the notation (4.6.10), the system (4.6.9) can be written more succinctly as

$$\dot{\bar{z}} = \bar{A}\bar{z} + \bar{B}\bar{v} + \bar{M}\bar{\mu}, \tag{4.6.11a}$$

$$0 = \bar{C}\bar{z} + \bar{N}\bar{\mu}. \tag{4.6.11b}$$

The new disturbance measure is

$$\bar{\mathcal{S}}_i(\bar{\mu}_i) = \bar{z}_i(0)^T \bar{Q}_i \bar{z}_i(0) + \int_0^h \|\bar{\mu}_i\|^2 dt = \mathcal{S}_i(\phi_i, \nu_i). \tag{4.6.12}$$

With the exception of the additional boundary conditions (4.6.7b), system (4.6.9) is the same system that we studied previously in this chapter with the same type of noise. However, the matrices can now be quite large. If one is going to use Riccati equations, there can be a severe computational penalty. The Riccati equation is a matrix equation so that the number of variables will increase as κ^2. Then one has to integrate these Riccati equations. Depending on whether one is using explicit or implicit integrators, the computational effort is proportional to the dimension or a power of the dimension. Since the work also, loosely, scales with the interval length, we get at best that the computational effort scales with κ^2. For even modestly sized systems, and for reasonable values of h, this can become excessive as the delay gets small.

There is a class of methods for which the numerical work for a given T and set of models is essentially independent of h for a wide range of h. These are sparse direct transcription algorithms like SOCS.

Thus if we are working with a moderate value of κ and using direct transcription software, the optimization software does not do significantly more work. Using the formulation of the models and applying the algorithm above, the optimizer finds one mesh on the interval $[0, h]$. This corresponds to a mesh on $[0, T]$ that is h periodic. The method of steps formulation causes significantly more computational work only when h is so small that the h-periodic mesh on $[0, T]$ is significantly finer than the mesh needed on $[0, T]$ to resolve the dynamics. The one exception to these comments about efficiency is that the delay can impact on the bandwidth of the sparsity patterns in the linear algebra routines used by the optimizer.

Algorithm Using Method of Steps

The algorithm for finding the minimum proper auxiliary signal now proceeds as described earlier in this chapter. The key difference from these earlier sections, in addition to problem size, is that now the state x satisfies a more complex set of boundary conditions. This in turn alters the boundary conditions on the λ variables in the optimization formulation. Suppose the additional boundary conditions are (4.6.7b). As we saw earlier in this chapter, the optimization problem for finding the auxiliary signal v can be formulated in two ways. Let $\bar{\lambda}$ be the same size as $\bar{x}$ with block elements λ_i. Then one optimization formulation is

$$(\gamma^*)^{-2} = \min_{\bar{v}} \delta^2(\bar{v}) \tag{4.6.13a}$$

subject to the constraints

$$\dot{\bar{z}} = \bar{A}\bar{z} + \bar{B}\bar{v} + \bar{M}\bar{\mu}, \tag{4.6.13b}$$

$$\dot{\bar{\lambda}} = -\bar{A}^T\bar{\lambda} + \bar{C}^T\bar{\eta}, \tag{4.6.13c}$$

$$\dot{\bar{\psi}} = F_0\bar{\psi} + G_0\bar{v}, \tag{4.6.13d}$$

$$\psi_0(0) = 0, \psi_i(0) = \psi_{i-1}(h), \ i = 1, \ldots, \kappa \tag{4.6.13e}$$

$$0 = C\bar{z} + \bar{N}\bar{\mu}, \tag{4.6.13f}$$

$$0 = V_\beta\bar{\mu} - \bar{M}^T\bar{\lambda} + \bar{N}^T\bar{\eta}, \tag{4.6.13g}$$

$$\dot{s} = \frac{1}{2}\bar{\mu}^T V_\beta\bar{\mu}, \quad s(0) = z_0(0)^T \hat{P}_\beta z_0(0) \tag{4.6.13h}$$

$$1 \leq s(h), \tag{4.6.13i}$$

$$\lambda_0(0) = -\hat{P}_\beta z_0(0), \quad \lambda_\kappa(h) = 0 \tag{4.6.13j}$$

$$z_j(0) = z_{j-1}(h), \quad j = 1, \ldots, \kappa, \tag{4.6.13k}$$

$$\lambda_j(0) = \lambda_{j-1}(h), \quad j = 1, \ldots, \kappa, \tag{4.6.13l}$$

$$0 < \beta < 1. \tag{4.6.13m}$$

Here $\frac{1}{2}\bar{\mu}^T V_\beta\bar{\mu} = \beta\|\bar{\mu}_0\|^2 + (1 - \beta)\|\bar{\mu}_1\|^2$.

4.6.2 Integral Terms

A number of variations on the delay systems (4.6.1) can be handled with the same approach. We mention just one. Some control problems have both point delays and distributed delays. In these cases the models (4.6.4) and (4.6.5) would have their first equation replaced by

$$\dot{x}(t) = Ax(t) + Gx(t - h) + \int_{-h}^{0} Ex(t + \tau)d\tau$$

$$+ Bv(t) + \int_{-h}^{0} Fv(t + \tau)d\tau + M\nu(t), \tag{4.6.14}$$

where A, G, B, M can depend on t and E and F can depend on τ. Even if the original problem does not have these integral terms, feedback terms often have integrals in them. Within the method of steps we can include these types of systems by the introduction of additional variables. For $j > 0$ we have that (4.6.14) is

$$\dot{x}_j = Ax_j + Gx_{j-1} + E(z_j + w_{j-1}) + Bv_j + F(u_j + r_{j-1}) + Mv_j, \quad (4.6.15a)$$

$$\dot{z}_j = x_j, \quad z_j(0) = 0, \quad (4.6.15b)$$

$$\dot{w}_j = x_j, \quad w_j(h) = 0, \quad (4.6.15c)$$

$$\dot{u}_j = v_j, \quad u_j(0) = 0, \quad (4.6.15d)$$

$$\dot{r}_j = v_j, \quad r_j(h) = 0. \quad (4.6.15e)$$

What we are doing is using for $jh < t < (j+1)h$ the fact that

$$\int_{-h}^{0} g(t+\tau)d\tau = \int_{t-h}^{t} g(s)ds = \int_{jh}^{t} g(s)ds + \int_{t-h}^{jh} g(s)ds. \quad (4.6.16)$$

Each of the integrals on the right can be written as a differential equation and a boundary condition. For $j = 0$ we must modify (4.6.15c) and (4.6.15e) as follows:

$$\dot{w}_0 = \phi, \quad w_0(h) = 0, \quad (4.6.17)$$

$$\dot{r}_0 = \psi. \quad (4.6.18)$$

The function ϕ is just the initial condition on the state. The ψ is more problematic. The function ψ is the auxiliary signal prior to time 0, which does not exist. When we are sending the auxiliary signal down the same channel as a preexisting control, we need to decide on ψ based on physical considerations. But in the case discussed here, it makes the most sense to just use $\psi = 0$.

The next example not only illustrates using the method of steps but also provides a specific example of the fact mentioned in chapter 3 that guaranteed failure detection may not be possible during normal operation without the use of an auxiliary signal.

Example 4.6.1 *We consider an example taken from the control of the Mach number in a wind tunnel [55]. Suppose that one model is*

$$\dot{x}_1(t) = -ax_1(t) + akx_2(t-h) + v_1, \quad (4.6.19a)$$

$$\dot{x}_2(t) = x_3(t) + v_2, \quad (4.6.19b)$$

$$\dot{x}_3(t) = -\delta^2 x_2(t) - 2\eta\delta x_3(t) + \delta^2(u(t) + v(t)) + v_3, \quad (4.6.19c)$$

$$y = x_1 + x_2 + v_4 \quad (4.6.19d)$$

where $a = (1.964)^{-1}, k = -0.0117, \delta = 6, \eta = 0.8,$ and $h = 1$. Note that the scaling of variables is set up so that for a constant control u there is a set point, or equivalently a constant solution, $x_1 = ku, x_2 = u, x_3 = 0$. Assume the detection horizon is $[0, \kappa h]$ and $Q_i = I, U = 0, W = 0$. The auxiliary signal is sent down the same channel as u. The failed system is the same as (4.6.19) except that $\delta = 4$ and $a = 1/2.892$. Neither δ nor a appears in the equations defining the set point, so that if u is constant and the state is at the set point, then a change in δ or a would not be detectable no matter how long the system is observed. Thus an auxiliary signal is needed to notice changes in δ or a.

Figure 4.6.1 gives the minimum energy v for $\kappa = 1$ when $u = 2$. The cost is $\|v\|^2 = 29.69814$. Since a constant u gives the same set point in both models, the

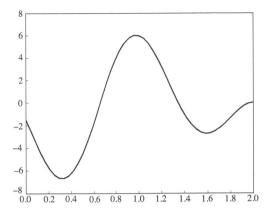

Figure 4.6.1 Minimum proper auxiliary signal for example 4.6.1.

*minimum energy auxiliary signal for this problem is independent of u for constant u.
In computing this answer, it was found that limiting β to a smaller range was helpful
in attaining a feasible solution on the initial coarse grids, as was eliminating some
of the constraints and thereby reducing the number of noise variables.*

4.6.3 General Delay Systems and Their Approximation

We now turn to considering more general problems with several delays and where
changes in delays may be part of the failure. We assume that we have m possible
models of the form

$$\dot{x}_i(t) = A_i x_i(t) + \sum_{j=1}^{r_i} G_{ij} x_i(t - \tau_{i,j}) + B_i v(t) + M_i \nu_i(t), \quad (4.6.20\text{a})$$

$$y(t) = C_i x_i(t) + N_i \nu_i(t), \quad (4.6.20\text{b})$$

$$x_i(t) = \phi_i(t), \; -h_i \le t < 0, \; x_i(0) = x_{i,0} \quad (4.6.20\text{c})$$

for $i = 0, \ldots, m - 1$, where v is an auxiliary signal to be determined, y is the
observed output, and $h_i = \tau_{i,r} > \tau_{i,r-1} > \cdots \tau_{i,1} > 0$. Let $h = \max_i h_i$. The
delays and system matrices may both vary from model to model. For simplicity of
discussion we shall take the system matrices as constants. However, the discussion
goes through with only the obvious changes if the system matrices are continuous
functions of t. We do not know ϕ_i and thus it is a type of disturbance or noise as
are the ν_i. Finally, we suppose the N_i are full row rank and the detection horizon is
$[0, T]$ with $T > h$. The two initial conditions $x_{i,0}$ and ϕ_i are considered to be two
independent disturbances. That is, there is no relationship between them other than
(4.6.21). We suppose that the disturbances $\phi_i, \nu_i, x_i(0)$ satisfy

$$S_i(\phi_i, \nu_i, x_i(0)) = \int_{-h_i}^{0} |\phi_i|^2 dt + x_i(0)^T Q_i x_i(0) + \int_{0}^{T} |\nu_i|^2 dt \le 1, \quad (4.6.21)$$

where Q_i is positive semidefinite. No weight is necessary on the second integral
since such a weight can be accommodated by redefining the N_i, M_i.

Let $\mathcal{A}_i(v)$ be the set of outputs y for (4.6.20) given that (4.6.21) holds. Then an auxiliary signal v is proper if it leads to disjoint output sets $\mathcal{A}_i(v)$. That is, $\mathcal{A}_i(v) \cap \mathcal{A}_j(v) = \emptyset$ if $i \neq j$. Notice that while the systems and states are infinite dimensional the output sets, y, and v are all the same as before, namely, L^2 signals with values in a finite-dimensional vector space.

To directly attack the auxiliary design problems for delay systems would require using the theory of optimal control of delay systems which can get quite technical and is well outside the scope of this book. Rather, we shall approximate the delay models by models without delays and apply the approach of the earlier sections to these models. In this approach, the inner minimization is replaced by its necessary conditions to create a new optimization problem for v. These necessary conditions turn out to have implications for our approximations, as will be shown later. In particular, their correct application requires small, but numerically critical, changes in the approximations.

4.6.3.1 Needed Theory

We first make a few observations about the auxiliary signal design problem for delay systems.

Let $\mathcal{L}g$ be the solution of the delay equation

$$\dot{x}_i(t) = A_i x_i(t) + \sum_{j=1}^{r_i} G_{ij} x_i(t - \tau_{i,j}) + g(t), \qquad (4.6.22a)$$

$$x_i(t) = 0, \text{ for } -h_i \leq t \leq 0. \qquad (4.6.22b)$$

Let $\Gamma\phi$ be the solution of

$$\dot{x}_i(t) = A_i x_i(t) + \sum_{j=1}^{r_i} G_{ij} x_i(t - \tau_{i,j}), \qquad (4.6.23a)$$

$$x_i(t) = \phi(t), \text{ for } -h_i \leq t < 0. \qquad (4.6.23b)$$

Finally, let Θ be the solution of

$$\dot{x}_i(t) = A_i x_i(t) + \sum_{j=1}^{r_i} G_{ij} x_i(t - \tau_{i,j}), \qquad (4.6.24a)$$

$$x_i(t) = 0, \text{ for } -h_i \leq t < 0, \ x_i(0) = I. \qquad (4.6.24b)$$

Then the output in (4.6.20) can be written as

$$y = C_i \mathcal{L}_i B_i v + C_i \mathcal{L}_i M_i \nu_i + C_i \Gamma_i \phi_i + \Theta_i x_{i,0} + N_i \nu_i. \qquad (4.6.25)$$

The various operators are all bounded linear operators from different L^2 spaces into the same L^2. Thus $\mathcal{A}_i(v)$ is a convex set in L^2 translated by the vector $C_i \mathcal{L}_i B_i v$. We have that the set of ϕ_i, ν_i satisfying (4.6.21) are L^2 bounded. If $Q_i > 0$, then the convex set $\mathcal{A}_i(v)$ is bounded. If some or all Q_i are indefinite, then $\Theta_i x_{i,0}$ is a finite-dimensional subspace. Thus we have the following lemma.

Lemma 4.6.1 *Suppose that $Q_i > 0$ for $i = 1, \ldots, m - 1$. Then cv is proper for sufficiently large scalars c if and only if $C_i \mathcal{L}_i B_i v \neq C_j \mathcal{L}_j B_j v$ for $i \neq j$.*

Proof. Notice that $C_i \mathcal{L}_i B_i v \neq C_j \mathcal{L}_j B_j v$ is all that is required for a multiple of v to be proper for a comparison of model i to model j. But if v is proper, then any larger multiple is also proper.   □

The method of steps does not work well, or at all, if there are multiple delays. It also leads to very-high-dimensional problems if h is small relative to T. We wish to consider the problems where not only are there multiple delays but a change in the delay itself could be a source of failure. Suppose then that we have one of the models of the form (4.6.20). We shall temporarily suppress the subscript i to simplify the notation and describe the approximation process for one model. We can rewrite (4.6.20) as follows. Let $U(t, s) = x(t + s)$ for $0 \leq t \leq T, -h \leq s \leq 0$. Thus for a fixed t, $U(t, s)$ is a function in L^2 which is x on the interval $[t - h, t]$. To guarantee that U is the function we want we require U to satisfy the following well known partial differential equation (PDE):

$$U_t(t, s) = U_s(t, s), \tag{4.6.26a}$$

$$U_t(t, 0) = AU(t, 0) + \sum_{j=1}^{r} G_j U(t, -\tau_j) + Bv(t) + Mv(t), \tag{4.6.26b}$$

$$U(0, s) = \phi(s), \quad -h \leq s < 0, \tag{4.6.26c}$$

$$U(0, 0) = x_0, \tag{4.6.26d}$$

$$y(t) = CU(t, 0) + Nv(t). \tag{4.6.26e}$$

There are a number of ways to approximate PDEs by ODEs. When applied to (4.6.26) we get an ODE model identification problem similar to that described in section 2. However, there are a number of technical problems to be resolved, which we do not discuss here, including the accuracy of the approximation, the relationship between optimization of the approximation and optimization of the original problem, and how to guarantee detection of the original problem when using approximate problems to compute the auxiliary signal. The interested reader is referred to [32].

We now turn to considering two types of approximations. Different methods of approximation can lead to different models. It is desirable to use the same approximation procedure in all models. Otherwise, there is the danger that the auxiliary signal will be designed to detect differences due to the approximation methods rather than differences in the original models.

4.6.3.2 Use of Differences

One method of approximating the PDE is the use of differences and the method of lines. This approach proceeds as follows. We pick a mesh for $[-h, 0]$ of the form

$$-h = s_0 < s_1 < \cdots < s_\rho = 0.$$

The one restriction on this mesh is that each $-\tau_j$ has to be a mesh point s_{m_j}. We suppose then that $-\tau_j = s_{m_j}$. The value of h and the delays may vary from model to model. Let

$$U_k(t) = U(t, s_k).$$

Then we get the following system of ordinary differential equations:

$$\dot{U}_0(t) = \sum_{j=0}^{2} \alpha_{0,j} U_h(t), \tag{4.6.27a}$$

$$\dot{U}_k(t) = \sum_{j=-1}^{1} \alpha_{k,j} U_{k+h}(t), \ 0 \le k < \rho - 1, \tag{4.6.27b}$$

$$\dot{U}_\rho(t) = AU_\rho(t) + \sum_{j=1}^{r} G_j U_{m_j}(t) + Bv(t) + M\nu(t), \tag{4.6.27c}$$

$$U_k(0) = \phi(s_k), \quad k < \rho, \tag{4.6.27d}$$

$$U_\rho(0) = x_0 \tag{4.6.27e}$$

where (4.6.27b) and (4.6.27a) come from (4.6.26a) while (4.6.27c) comes from (4.6.26b). If we have $0 < k < \rho$ and let $\delta = s_{k+1} - s_k$, $\epsilon = s_k - s_{k-1}$, then we can take

$$\alpha_{k,1} = \frac{\epsilon}{\delta(\epsilon + \delta)}, \quad \alpha_{k,-1} = -\frac{\delta}{\epsilon(\epsilon + \delta)}, \tag{4.6.28}$$

$$\alpha_{k,0} = \frac{\epsilon}{\delta(\epsilon + \delta)} - \frac{\delta}{\epsilon(\epsilon + \delta)}. \tag{4.6.29}$$

This provides an approximation for the spatial derivative which is $O(\delta\epsilon)$. To get a similar accuracy for the U_0 equation we need to use a one-sided approximation using two extra values. Let $\delta = s_1 - s_0$, $\epsilon = s_2 - s_1$. We can then take

$$\alpha_{0,0} = -\alpha_{0,2} - \alpha_{0,1}, \quad \alpha_{0,1} = \frac{\delta + \epsilon}{\delta\epsilon}, \quad \alpha_{0,2} = -\frac{\delta}{\epsilon(\delta + \epsilon)}. \tag{4.6.30}$$

Two things should be noticed. First we have that

$$a_{k,1} \le \frac{1}{\delta}, \quad a_{k,-1} \le \frac{1}{\epsilon}$$

so that the coefficients grow only as the reciprocal of the mesh size. Secondly if $\epsilon = \delta$, then (4.6.29) simplifies to

$$a_{k,1} = \frac{1}{2\delta}, \quad a_{k,0} = 0, \quad a_{k,-1} = -\frac{1}{2\delta}.$$

This procedure has transformed the delay model equation to an approximate ordinary differential equation model. We also must transform the first term in the noise measure (4.6.21). Let $U_{i,j}$ be the j vector in the approximation (4.6.27) for model i. Let $\gamma_{i,j}$ be the collocation coefficients for approximating integrals on $[-h_i, 0]$ using the grid points $s_{i,j}$. Then (4.6.21) becomes

$$\tilde{S}_i(\phi_i, \nu_i, x_{i,0}) = \sum_{j=0}^{\rho_i - 1} \gamma_{i,j} |U_{i,j}(0)|^2 + \gamma_{i,\rho_i} |\phi_i(s_{\rho_i})|^2$$

$$+ U_{i,\rho_i}(0)^T Q_i U_{i,\rho_i}(0) + \int_0^T |\nu_i|^2 dt \le 1. \tag{4.6.31}$$

Note that (4.6.31) is the same type of noise measure studied previously for the problem without delays present. However, there is one important difference. In the optimization approach we use here, a maximization-minimization problem is solved by replacing the minimization by the necessary conditions for the minimization along with a quantity that returns the value of the minimization. In the case without delays this was a boundary value problem. However, in general, if the cost has a nonnegative term that does not appear in any dynamic equations or constraints, the necessary conditions imply that this term is zero. This adds an extra equation to the necessary conditions and hence an extra equation to the optimization formulation. For our problem here, note that $\phi_i(0)$ appears only in the noise bound (4.6.31) and nowhere in the dynamics or initial conditions in (4.6.27). Accordingly, we have an additional necessary condition that

$$\phi_i(0) = 0. \tag{4.6.32}$$

Thus instead of (4.6.31) we get

$$\widetilde{S}_i(\phi_i, \nu_i, x_{i,0}) = \sum_{j=0}^{\rho_i - 1} \gamma_{i,j} |U_{i,j}(0)|^2 + U_{i,\rho_i}(0)^T Q_i U_{i,\rho_i}(0) + \int_0^T |\nu_i|^2 dt. \tag{4.6.33}$$

It is easy to see that if the condition (4.6.32) is not added, there is an incorrect solution of the optimization problem which has $v = 0$. We have seen this solution occur in computational tests which did not assume (4.6.32).

We shall use collocation schemes for which all the $\gamma_{i,j} > 0$. For example, if we are using a trapezoidal approximation, we have (suppressing the i subscript on all terms)

$$\gamma_0 = \frac{s_1 - s_0}{2}, \quad \gamma_\rho = \frac{s_\rho - s_{\rho-1}}{2},$$

$$\gamma_j = \frac{s_{j+1} - s_{j-1}}{2} \text{ if } j \notin \{0, \rho\}. \tag{4.6.34}$$

4.6.3.3 Spline Approximation

An alternative approach to replacing the delay system with an ODE approximation is by approximating the various operators and functions from finite-dimensional subspaces. The approach we consider is from [45] and we follow their notation. We shall leave out much of the technical detail. It should be noted that this approach allows us to consider more complex models than (4.6.20a). In particular, one can include terms such as $\int_{-h_i}^0 F_i(\theta) x_i(t + \theta) d\theta$.

Pick N to be the partition size. Let t_k^N be a partition of $[-h, 0]$, $k = 0, \ldots, N$. Both ends are doubled by adding t_{-1}^N, t_{N+1}^N. The point t_0^N starts at the right end of $[-h, 0]$. Let $B_k^N(\theta)$ be the usual "hat" function. That is, B_k^N is piecewise linear and zero at all mesh points except t_k^N, where it is 1. Thus if $g(t)$ is a function on $[-h, 0]$, we have that $\sum_{j=0}^N g(t_k^N) B_k^N(t)$ is a piecewise linear interpolation of $g(t)$ at the grid points. It is assumed that every delay of the original system falls into at most one of the subintervals. Note that, unlike with differences, the grid points do not have to include the delay so we can use a uniform grid.

If the initial data are in $R^n \times L^2$, and $u \in L^2$, then solutions of (4.6.20a) are in both $L^2(-h, T; R^n)$ and $H^1(0, T; R^n)$. (H^1 consists of continuous functions with L^1 derivatives.) We take $\text{dom}(\mathcal{A}) = \{(\eta, \phi) \in Z | \phi \in H^1(-h, 0; R^n), \eta = \phi(0)\}$ and $\mathcal{A}(\phi) = \sum A_i \phi(\theta_i)$. H^1 is given the norm and inner product

$$\langle \phi, \psi \rangle_{H^1} = \langle \phi(0), \psi(0)i \rangle_{R^n} + \left\langle \dot{\phi}, \dot{\psi}i \right\rangle_{L^2}.$$

We define Ψ as an isomorphism between $\text{dom}(\mathcal{A})$ and H^1 by $\Psi(\phi(0), \phi) = \phi$. Then we have $\Psi^{-1}(\phi) = (\phi(0), \phi)$. Other notation includes $E_k^N = \aleph_{[t_k^N, t_{k-1}^N)}$, $k = 1, \ldots, N$, for the characteristic function of a subinterval; $\widehat{E}_0^N = (I_n, 0)$ for R^n; $\widehat{E}_k^N = (0, I_n)$ for the past; $W^N = \text{span}\{E_k^N I_n : 1 \le k \le N\}$ for the approximation subspace of $L^2[-r, 0]$; $Z^n = R^n \times W^N = \text{span}\{E_k^N I_n : 0 \le k \le N\}$ for the approximation subspace of Z; $X^N = \text{span}\{B_k^N I_n : 0 \le k \le N\}$ for the approximation subspace of H^1; and $Z_1^N = \Psi^{-1} X^N$ for the approximation subspace of $\text{dom}(\mathcal{A})$. In addition we have the basis matrices $E^N = (E_1^N I \cdots E_N^N I)$, $\widehat{E}^N = (\widehat{E}_0^N \cdots \widehat{E}_N^N)$, and $\widehat{B}^N = (B_0^N I \cdots B_N^N I)$.

Then $z = (\eta, \phi) \in Z^N$ can be written $(\eta, E^N a^N) = \widehat{E}^N \text{col}(\eta, a^N)$ where a^N is a coordinate vector of $\phi \in W^N$ with respect to the basis. Similarly, if $\phi \in X^N$, then $\phi = B^N b^N$, $b^N = \text{col}(b_0^N, \ldots, b_N^N)$.

Let

$$Q^N = \begin{pmatrix} 1 & 0 & \cdots & \cdots & 0 \\ \frac{1}{2} & \frac{1}{2} & \ddots & \cdot & \vdots \\ 0 & \ddots & \ddots & \ddots & \vdots \\ \vdots & \ddots & \ddots & \ddots & 0 \\ 0 & \cdots & 0 & \frac{1}{2} & \frac{1}{2} \end{pmatrix} \otimes I_n \tag{4.6.35}$$

where the first matrix is $(N+1) \times (N+1)$. For $k = 0, \ldots, N$, let

$$D_k^N = L(B_k^N) = \sum_{j=0}^{c} A_j B_k^N(\theta_j). \tag{4.6.36}$$

Also let

$$H^N = \begin{pmatrix} D_0^N & \cdots & D_N^N \\ 0 & \cdots & 0 \end{pmatrix} + \frac{N}{r} \begin{pmatrix} 0 & \cdots & \cdots & \cdots & 0 \\ 1 & -1 & 0 & \cdots & 0 \\ 0 & \ddots & \ddots & \ddots & \vdots \\ \vdots & \ddots & \ddots & \ddots & 0 \\ 0 & \cdots & 0 & 1 & -1 \end{pmatrix} \otimes I_n, \tag{4.6.37}$$

where the first matrix in the Kronecker product is $(N+1) \times (N+1)$ and the first matrix after the equality has lots of zero rows. Then

$$(Q^N)^{-1} = \begin{pmatrix} 1 & 0 & \cdots & \cdots & 0 \\ -1 & 2 & \ddots & \cdot & \vdots \\ 1 & -2 & 2 & \ddots & \vdots \\ \vdots & \ddots & \ddots & \ddots & 0 \\ (-1)^{N+1} & \cdots & 2 & -2 & 2 \end{pmatrix} \otimes I_n. \qquad (4.6.38)$$

B^N and M^N are block column matrices whose only nonzero entries are the first ones, which are B and M, respectively. Similarly, C^N and N^N are block row matrices whose only nonzero entries are the first ones, which are C and N, respectively.
    Let

$$A^N = H^N (Q^N)^{-1}. \qquad (4.6.39)$$

The delay differential equation is then approximated by the ordinary differential equation

$$\frac{d}{dt}(a^N)(t) = A^N a^N(t) + B^N v(t) + M^N \eta(t), \qquad (4.6.40a)$$

$$y(t) = C^N a^N(t) + N^N \eta(t). \qquad (4.6.40b)$$

4.6.3.4 Computational Examples

We shall now work several computational examples involving delays. In addition to showing the usefulness of the approach given here, they will serve to illustrate some theoretical points. It suffices to consider two models of the form

$$\dot{x}_i(t) = A_i x_i(t) + G_i x_i(t - h) + B_i v + M_i \nu_{i,1}, \qquad (4.6.41a)$$

$$y = C_i x_i(t) + N_i \nu_{i,2}. \qquad (4.6.41b)$$

Example 4.6.2 *Our first example (4.6.42) will be used to illustrate a number of points. It is chosen since it is possible to reformulate this problem using the method of steps and obtain the exact solution for comparison purposes. The detection horizon is $[0, 2]$. We take $Q_i = I$ and $W = 0, U = 0$.*

$$\dot{x}_0 = -2x_0 + x_0(t - 1) + v + \nu_1, \qquad (4.6.42a)$$

$$y = x_0 + \nu_2, \qquad (4.6.42b)$$

$$\dot{x}_1 = -3x_1 + x_1(t - 1) + v + \nu_3, \qquad (4.6.42c)$$

$$y = x_1 + \nu_4. \qquad (4.6.42d)$$

    Figure 4.6.2 shows the "true" minimum energy auxiliary signal v computed using steps and the auxiliary signals v_ρ obtained using differences with several values of ρ where ρ is the number of spatial grid points used in (4.6.27).
    Observe that the approximation appears better for increasing ρ although on this problem the coarsest approximation using $\rho = 4$ already did quite well. The first column of table 4.6.1 gives the norm of the auxiliary signal.

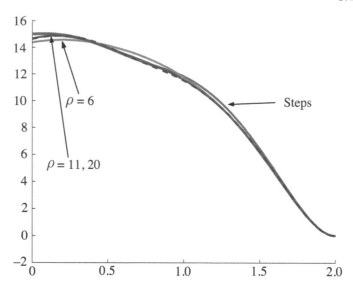

Figure 4.6.2   Minimum energy auxiliary signal v and difference
approximations v_6, v_{11}, v_{20} for example 4.6.2.

Table 4.6.1 $\|v\|, \|v_\rho\|, c_\rho$ for example 4.6.2 using differences with $\rho = 4, 6, 11$.

	$\|v_\rho\|$	c_ρ	$\|c_\rho v_\rho\|$
v_4	15.37964	1.00063	15.38943
v_6	15.50291	0.99234	15.38421
v_{11}	15.45982	0.99479	15.37929
v	15.33738	1	15.33738

The question arises as to how good is the minimum energy auxiliary signal from the approximation. To examine this we took v_ρ and found the minimum value of c_ρ such that $c_\rho v_\rho$ was proper in the original system. This computation showed that, for this example, not only were the v_ρ proper for the true problem, but $c_\rho v_\rho$ was close to the minimum. Table 4.6.1 shows the norms of v, v_ρ, and $c_\rho v_\rho$ for several values of ρ.

We have also solved this problem using the spline approximations described earlier. The results are shown in figure 4.6.3.

It is interesting to note that as ρ and N are increased the difference and spline approximations for v seem to converge to the same function. In fact, by $\rho = N = 20$ they are almost identical, as shown in figure 4.6.4.

The spline and difference approximation minimum energy auxiliary signals differ slightly from those of the steps solution. However, the norm of the spline approximation is very close to that of the step solution as seen in table 4.6.2.

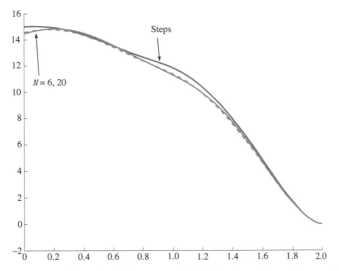

Figure 4.6.3 Auxiliary signals for example 4.6.42 using splines for several N.

Table 4.6.2 Norm of spline approximation solutions for v for example 4.6.42.

N	$\|v_N\|$	Relative Error
3	15.49152	0.0129546
4	15.40969	0.0076039
5	15.37202	0.0051408
20	15.301224	0.0005113
v	15.2934	0

We see that both the splines and differences do an excellent job of approximating and give us high-quality auxiliary signals using low values of N and ρ. The difference in norm between the approximate v and the steps v is quite small and the relative error shown in table 4.6.2 is around 10^{-5}. This suggests that the difference is due to the fact that the answers from all three approaches are essentially equal in terms of the optimization problem, and the remaining difference is due to how they approach the minimum. Graphically, the spline and difference approximations appear to be equally good. The tables show that the spline approximations give a slightly better estimate of $\|v\|$.

Example 4.6.3 *This example illustrates a computational subtlety. We take $Q_i = I$ and $U = 0, W = 0$. Suppose that the two models are*

$$\dot{x}_0 = -2x_0 + x_0(t-1) + v + \nu_1, \qquad (4.6.43a)$$

$$y = x_0 + \nu_2, \qquad (4.6.43b)$$

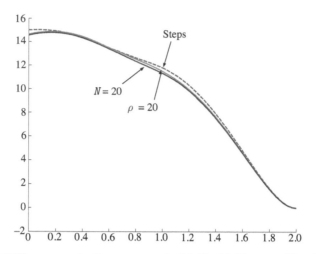

Figure 4.6.4 Differences and splines on example 4.6.42 with $N = \rho = 20$ and the steps v.

and

$$\dot{x}_1 = -2x_1 + x_1(t - 0.4) + v + \nu_3, \qquad (4.6.43c)$$

$$y = x_1 + \nu_4, \qquad (4.6.43d)$$

so that the failure consists of a change in the delay interval.

Suppose that we decide for convenience to use a delay interval of $[-1, 0]$ for both models. When we solve the problem for $\rho = 5$ or $N = 5$ we get the auxiliary signal in figure 4.6.5.

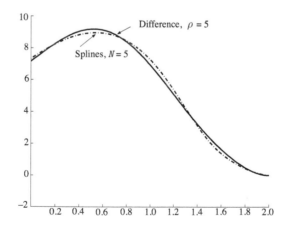

Figure 4.6.5   Minimum energy auxiliary signal for example 4.6.3 using splines and differences with $h_1 = h_2 = 1$ and $N = \rho = 5$.

However, if we take $\rho \geq 11$ we begin to experience numerical difficulties. To understand what is happening we note that the approximate problem is a reasonable

finite-dimensional model. However, in the continuous problem the initial perturbation for model 1 on $[-1, -0.4]$ appears in the noise measure. Thus the correct
formulation of the optimization version of our approach would require the initial
perturbation to be zero on this interval. The failure to take this difference into account leads to a numerical breakdown as the width of the mesh used to partition
the delay interval goes to zero. Thus, as with our handling of $\phi(0)$, one must keep
the theoretical approach in mind and make sure that all the necessary conditions are
included in the formulation.

Example 4.6.4 *Suppose that two delays are present and that the two models are
given below. Again we take* $Q_i = I, W = 0, U = 0.$

$$\dot{x}_0 = -2x_0 + x_0(t - 1) + v + \nu_1, \quad \cdot \tag{4.6.44a}$$

$$y = x_0 + \nu_2, \tag{4.6.44b}$$

$$\dot{x}_1 = -2x_1 + 0.5x_1(t - 1) + 0.5x_1(t - 0.4) + v + \nu_3, \tag{4.6.44c}$$

$$y = x_1 + \nu_4. \tag{4.6.44d}$$

*Here the failure consists of an additional shorter delay of the state. This type of
change can occur if the delay in a feedback loop is altered, for example by failure of
the hardware or software implementing the feedback. The minimum energy auxiliary
signal found using difference approximations is given in figure 4.6.6.*

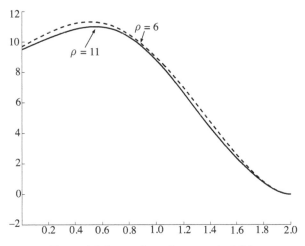

Figure 4.6.6 v_6 and v_{11} for example 4.6.4.

The spline and difference solutions of example 4.6.4 are almost identical as seen
in figure 4.6.7. It is interesting to note the difference in shape between the auxiliary
signals in figure 4.6.6 and those in figure 4.6.2. In example 4.6.4, where part of the
difference in the two models is due to the additional delay of 0.4, we see that v acts
most strongly in the interval $[0.4, 0.6]$.

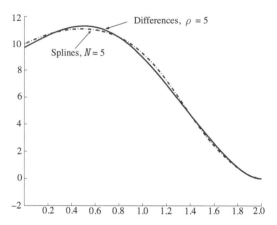

Figure 4.6.7 Spline and difference approximations with $N = \rho = 5$ on example 4.6.4.

4.7 SETTING ERROR BOUNDS

Up until now we have always taken the error bound to be equal to 1 since the problem can be rescaled from the original bound d. However, in an actual application using the techniques of either this chapter or chapter 3, d must be chosen. Care must be exercised in choosing d. An incorrect d can make the resulting test either too conservative or too lax. This section describes one way to set the error bounds. It is similar to the suggestion of Petersen and MacFarlane [71]. We illustrate by considering the additive noise case but the model uncertainty case is handled in the same way.

Suppose that we have a model 0 for the normal system. This model can be derived from physical principles or from a numerical procedure. In any event, we suppose that model 0 is

$$\dot{x}(t) = Ax(t) + Bv(t) + Ru(t) + M\nu(t), \qquad (4.7.1a)$$
$$y(t) = Cx(t) + Dv(t) + N\nu(t), \qquad (4.7.1b)$$

where v is an auxiliary signal to be determined and u is another known signal.

Detection is to be carried out over an interval $[0, T]$. The initial condition $x(0)$ is unknown and treated as another disturbance. We suppose that the disturbances $\nu, x(0)$ are to satisfy a bound

$$\mathcal{S}_i(\nu, x(0)) = x(0)^T Px(0) + \int_0^T |\nu(t)|^2 dt \le d. \qquad (4.7.2)$$

The problem is that of finding d. Once d is found, we adjust P, M, N to make $d = 1$.

We suppose that r experiments have been run over the horizon $[0, T]$ on the normal system so that we have the observations $\{(y^i, u^i, v^i) : i = 1, \dots, r\}$. If the data are discrete, we replace them with interpolating splines so that we have functions for the (y^i, u^i, v^i). We may or may not have values for $x^i(0)$.

Depending on the application there are different concerns in setting d. Let us suppose that we want to have few false failures. That is, we do not want to think there is a failure when, in fact, the behavior is consistent with the model.

Thus we look for a bound d that is as small as possible and yet does not cause us to decide there has been a failure based on an observation that we know can come from model 0. That is,

$$d = \max_i \min_{\nu, x(0)} x(0)^T P x(0) + \int_0^T |\nu(t)|^2 dt, \qquad (4.7.3)$$

$$\dot{x}(t) = Ax(t) + Bv^i(t) + Ru^i(t) + M\nu(t), \qquad (4.7.4a)$$

$$y^i(t) = Cx(t) + Dv^i(t) + N\nu(t), \qquad (4.7.4b)$$

where the inner minimization of (4.7.3) is subject to the constraints (4.7.4). If there are estimates $x^i(0)$ for $x(0)$, then they are used in place of the variable $x(0)$. In order to simplify the exposition we will assume that $x(0)$ is not measured.

The outer maximization is over a finite set of numbers and is easily computed, so we focus on the inner minimum. Since we are interested only in the value of d we may perform an orthogonal transformation once on ν and get

$$\dot{x} = Ax + Bv^i + Ru^i + M_1\nu_1 + M_1\nu_2, \qquad (4.7.5a)$$

$$y^i = Cx + Dv^i + N\nu_1, \qquad (4.7.5b)$$

where N is now invertible. Eliminating ν_1, we get the unconstrained optimal control problem

$$\min_{\nu, x(0)} x(0)^T P x(0) + \int_0^T |\nu_2|^2 + |N^{-1}(y^i - Cx - Dv^i)|^2 dt, \qquad (4.7.6a)$$

$$\dot{x} = (A - M_1 N^{-1} C)x + [(B - M_1 N^{-1}D)v^i + Ru^i] + M_2\nu_2. \quad (4.7.6b)$$

This is a quadratic regulator problem with a positive definite weight on the noise (control) and an exogenous signal $[(B - M_1 N^{-1}D)v^i + Ru^i]$. It can be solved by any of the usual methods. Maximizng the value of (4.7.6a) over i gives us a lower bound for the noise measure for model 0.

Where to set the noise bounds for the failure model may depend on design and safety issues. If we can perform experiments on a failed plant, then we can repeat the above procedure and get a value d_i for each of the failure models. If we do not want to miss a failure and all the data are from the failed model, then $\max_i$ in (4.7.3) is replaced by $\min_i$.

4.8 MODEL UNCERTAINTY

We conclude this chapter by discussing the formulation of model uncertainty directly as an optimization problem. We focus on the two-model case. Recall from chapter 3 that after combining the equations, and after some rewriting, we may assume that the two models take the form

$$\dot{x}_i = A_i x_i + B_i v + M_i \nu_i, \qquad (4.8.1a)$$

$$E_i y = C_i x_i + D_i v + N_i \nu_i. \qquad (4.8.1b)$$

In this new notation, the constraint (or noise measure) on the initial condition and uncertainties (2.3.24) can be expressed as

$$S_i(v, s) = x_i(0)^T \hat{P}_i x_i(0) + \int_0^s v_i^T J_i v_i \, dt < 1, \ \forall \ s \in [0, T], \tag{4.8.2}$$

where the J_i's are signature matrices. As discussed in chapter 2, the bound (4.8.2) allows for both additive and model uncertainty.

We wish to find the proper v of smallest norm $\delta(v)$ given in (4.2.3a) with $\dot{\psi} = F_i \psi + G_i v$ for which there exists no solution to (4.8.1a), (4.8.1b), and (4.8.2) for $i = 0$ and 1 simultaneously.

As shown in chapter 3, the nonexistence of solutions to (4.8.1a), (4.8.1b), and (4.8.2) is equivalent to

$$\sigma(v, s) \geq 1 \text{ for some } 0 \leq s \leq T, \tag{4.8.3a}$$

where

$$\sigma(v, s) = \max_{\beta \in [0,1]} \phi_\beta(v, s) \tag{4.8.3b}$$

with

$$\phi_\beta(v, s) = \inf_{\substack{\nu_0, \nu_1, y \\ x_0, x_1}} \beta S_0(v, s) + (1 - \beta) S_1(v, s) \tag{4.8.3c}$$

subject to (4.8.1a),(4.8.1b), $i = 0, 1$. Notice that there are now two parameters s, β, whereas with only additive uncertainty we had only β. The parameter s also enters in a different way and creates some new computational issues.

Using the notation

$$x = \begin{pmatrix} x_0 \\ x_1 \end{pmatrix}, \ \nu = \begin{pmatrix} \nu_0 \\ \nu_1 \end{pmatrix}, \ A = \begin{pmatrix} A_0 & 0 \\ 0 & A_1 \end{pmatrix}, \ M = \begin{pmatrix} M_0 & 0 \\ 0 & M_1 \end{pmatrix},$$

$$D = F_0 D_0 + F_1 D_1, \ C = \begin{pmatrix} F_0 C_0 & F_1 C_1 \end{pmatrix}, \ N = \begin{pmatrix} F_0 N_0 & F_1 N_1 \end{pmatrix},$$

$$\hat{P}_\beta = \begin{pmatrix} \beta \hat{P}_0 & 0 \\ 0 & (1 - \beta) \hat{P}_1 \end{pmatrix}, \ J_\beta = \begin{pmatrix} \beta J_0 & 0 \\ 0 & (1 - \beta) J_1 \end{pmatrix}, \ B = \begin{pmatrix} B_0 \\ B_1 \end{pmatrix},$$

where

$$F = \begin{pmatrix} F_0 & F_1 \end{pmatrix} = \begin{pmatrix} E_0 \\ E_1 \end{pmatrix}^\perp, \tag{4.8.4}$$

we can reformulate the problem (4.8.3c) as follows:

$$\phi_\beta(v, s) = \inf_{\nu, x} x(0)^T \hat{P}_\beta x(0) + \int_0^s \nu^T J_\beta \nu \, dt \tag{4.8.5a}$$

subject to

$$\dot{x} = Ax + Bv + M\nu, \tag{4.8.5b}$$

$$0 = Cx + Dv + N\nu. \tag{4.8.5c}$$

The optimization problem for the minimum norm proper v thus is

$$\min_v \delta^2(v) \tag{4.8.6a}$$

subject to the constraints

$$\dot{x} = Ax + Bv + M\hat{v}, \tag{4.8.6b}$$

$$\dot{\psi} = F_2\psi + G_2v, \quad \psi(0) = 0, \tag{4.8.6c}$$

$$\dot{\lambda} = -A^T\lambda + C^T\eta, \quad \lambda(0) + \hat{P}_\beta x(0) = 0, \quad \lambda(s) = 0, \tag{4.8.6d}$$

$$0 = Cx + Dv + N\hat{v}, \tag{4.8.6e}$$

$$0 = 2J_\beta\hat{v} - M^T\lambda + N^T\eta, \tag{4.8.6f}$$

$$\dot{z} = \hat{v}^T J_\beta \hat{v}, \quad z(0) = x(0)^T \hat{P}_\beta x(0), \tag{4.8.6g}$$

$$0 < \beta < 1, \tag{4.8.6h}$$

$$0 \le s \le T, \tag{4.8.6i}$$

$$z(s) \ge 1. \tag{4.8.6j}$$

We know from the preceding discussion that when J_β is indefinite not all β will admit a finite solution. Our previous computational experience suggests that if estimates on β are available then bounding β more tightly can be helpful, especially if the optimization software starts on coarse grids.

Some care must be taken in how this problem is set up, since it is highly nonlinear in the parameter s and the problem can have some degenerate arcs. To illustrate, suppose that $U = 0, W = 0$. Suppose the detection horizon is $[0, T]$, $\hat{v}$ is the minimum norm v, and $\hat{s}, \hat{\beta}$ are the values of s, β for which $\phi_{\hat{\beta}}(\hat{v}, \hat{s}) = 1$. If $\hat{s} < T$, then $\hat{v}(t) = 0$ for $s \le t \le T$.

Depending on the software used there are a number of ways to handle s. Some software allows for multiple phases and the phase transition can be a variable. If the software requires fixed time intervals, then it is often possible to consider all functions in the form of $g(\alpha t)$, where α is a parameter, and thus implicitly allow for rescaling of the interval.

Chapter Five

Remaining Problems and Extensions

In this book we have presented a methodology for the design of auxiliary signals for use in active failure detection and isolation in dynamical systems. In particular, we have designed methods and developed the supporting theory for designing optimal auxiliary signals and associated on-line detection filters which can be used to detect and identify failures in real time. We have also considered computational aspects of the method and suggested a number of numerical algorithms. But there are a number of natural questions to ask that have not been addressed yet. Some of these problems were mentioned briefly in the text. In this chapter we will briefly discuss some of these remaining issues as well as some of the issues we see arising in the future with active failure detection.

Active failure detection will become increasingly important in the years ahead. There is an increasing move toward unmanned vehicles in space, under water, in the military, and in hazardous conditions. There is also a movement toward increased efficiency and smaller numbers of people involved in daily operations be it in manufacturing plants or in cargo ships at sea. In all of these areas it is becoming expensive or impractical to have a large number of skilled personnel regularly checking operations or to have regular extensive routine maintenance. Rather, there is a movement toward more elaborate sensors and algorithms to operate those sensors, which will decide when repair or replacement or switching to a backup system is needed prior to failure but no sooner than needed. This approach is sometimes referred to as *condition-based maintenance* or CBM.

Actual operations will likely be a mix of traditional scheduled maintenace and CBM with some maintenance being performed during scheduled downtimes but CBM being used to extend the operational period between scheduled checks.

In addition to its use as part of the primary failure detection system, the active approach has another important role to play in extended operations. Sensors, whether on an automotive engine or in a space vehicle, often operate in a harsh environment. One common type of failure is sensor failure. Upon sensor failure the previously designed failure detection algorithms, regardless of whether they are passive or active, may no longer work. This sensor failure can sometimes be compensated for by the use of an active approach to access and monitor the status of the system with a new auxiliary signal.

5.1 DIRECT EXTENSIONS

Some of the remaining open problems directly concern the same types of models that were considered in this book.

Optimization Goal

In this book the cost of an auxiliary signal was a combination of the size of the auxiliary signal and the size of various responses to that signal, such as deviation of the state from its normal operating point. There are a number of other natural optimization goals. For example, given a noise measure and a cost on the auxiliary signal, one could ask what is the shortest time interval over which perfect identification can occur given a specified size for the auxiliary signal. While we have not done so, this problem of minimizing the detection horizon given a signal size can be done using the techniques of this book. A γ iteration is replaced by a T iteration. Another variation on this type of problem is to ask, given a signal, if there is a noise bound for which it is proper, and if so, what is the largest such noise bound.

Construction of Uncertain Models

When constructing an uncertain model there are a number of choices that are made dealing both with the formulation of the noise bounds and how the size of the auxiliary signal is measured. Work needs to be done on how to best choose these weights. Different choices of weights will result in different minimal auxiliary signals. Suppose, for example, we have data from both a properly operating and a failure mode of a process. Given the data, how does one choose the various parameters in the noise bound so that the algorithms will compute the smallest possible auxiliary signal that guarantees identification and makes decisions consistent with the known data?

The uncertain models we use to model system behaviors are very general and capture various types of uncertainty. There is, however, a need for a systematic method for constructing model parameters for realistic applications. These model parameters include the models themselves, the matrices characterizing the uncertainty, the weights in the noise bounds, and the choice of cost. This is, of course, a problem that arises in many areas including control. But in our theory, where we use multiple models for the system and its failure modes, the tightness of the models is very important. If the models are chosen too conservatively, where here conservative means that they admit an unnecessarily large amount of noise and variation, then the search for an auxiliary signal may either fail or lead to an unnecessarily large, and perhaps impractical, auxiliary signal.

Often in failure detection applications the only way to construct models is by using identification techniques based on measurement data. That is particularly the case with the models associated with the failure modes. Most identification techniques have commonly used work with stochastic models and are not currently well adapted to our deterministic framework. There are some techniques used in other areas of control based on least squares approaches that might be useful in our setting. Studies are needed to find ways to directly construct, using measured data, the type of uncertain models we consider in this book.

Feedback

The auxiliary signals we have considered in this book are mostly for applications where the detection period, the time over which the auxiliary signal is to be applied, is short. They correspond to short, small, maneuvers for gently "shaking" the system in order to exhibit abnormal behaviors. These signals are computed off-line and applied as an open loop control. In some cases, there might be advantages in using feedback for generating the auxiliary signal. This problem has not been addressed in this book except in our discussions of early detection where we give a method for sometimes ending the test early.

The deterministic approach adopted here may or may not be amenable to this type of auxiliary signal generation. We have always considered the worst case. This is necessary for robustness but makes it difficult to construct a measure of efficiency for the feedback.

A related question arises when the large scale behavior of the system is nonlinear and it is desired to compute auxiliary signals. One approach could be to apply the techniques of the earlier chapters to linearized models and use them to compute an auxiliary signal. If there is a single operating point about which we do the linearization, then our previous discussion of small nonlinearities applies. However, if we assume that the system undergoes a wide range of behavior, then the situation is more complicated. One option is to assume that there are families of models for both the failed and unfailed systems, along with an estimate of the state. The state estimate would then tell you which models to use for the failed and unfailed systems. Then, adopting a gain scheduling approach, a library of auxiliary signals all computed off-line would be stored. In this case the choice of auxiliary signal would depend on the state estimate. For systems that undergo periodic communications with a base station, some of this library could be stored at the base station, saving on memory requirements.

Another alternative would be to move all, or part, of the computation of the auxiliary signal on-line. This would probably require less memory but more computational power. The on-line approach would require a reexamination of the algorithms and possibly the development of new ones geared to on-line computation of the auxiliary signal.

Coding for Online Use

In specific applications, memory or other considerations may make it necessary to consider specialized coding of the online signal. we have not considered this problem here.

5.2 HYBRID AND SAMPLED DATA SYSTEMS

In this book we have considered both discrete- and continuous-time systems. However, in many applications there is a mixture of both. For example, there could be continuously varying mechanical parts or chemical reactions and discrete events triggered by controllers or outside events. These systems are called hybrid systems. The modeling and control of hybrid systems is currently an area of active research. There is not really a general theory that covers all cases. If the discrete events happen infrequently, then it is usually not too difficult to model and simulate the hybrid system. But as the frequency of the events increases so do the difficulties. The design of auxiliary signals needs to be extended to hybrid systems.

A related problem concerns *sampled data systems*. With a sampled data system there is an underlying continuous process. However, it is sampled (measured) only at discrete times. Similarly, the control or auxiliary signal is updated only at discrete times. Depending on the application, the auxiliary signal could be piecewise constant or continuous. The design of auxiliary signals for sampled data systems is important for full utilization of the auxiliary signal approach, since many continuous systems are monitored and controlled with sampled data controllers. If the sampling time is short relative to the system dynamics, then the solution of the continuous-time problem can often be used with little loss in performance on the sampled data system. However, as the sampling rate becomes less frequent, an extension of the methodology of this book is needed. A first step is made in [64].

One can also have discrete processes which are sampled on a slower time scale than the time scale at which the discrete process operates.

5.3 RELATION TO STOCHASTIC MODELING

We have considered only deterministic models. There are some differences between the deterministic and stochastic approaches, but as we have noted, there are some similarities and connections. It is important to examine these interconnections more carefully. One possible outcome will be an improvement of some existing approaches.

The methodology developed in this book has been based on a multimodel design of the normal and failed systems using uncertain deterministic models. This has allowed us to deal with robustness issues in a natural way. It has also allowed formulation in terms of a mathematical optimization problem for which we have developed solutions.

In principle, our methodology is still usable even if we have stochastic models in either discrete time or continuous time representing normal and failed modes of the system. For that, the probabilistic information concerning system trajectories is used to divide the set of trajectories into different subsets corresponding to different modes. This separation is done based on the likelihood of each trajectory given the mode and leads to deterministic uncertain models to which our methodology applies.

Our approach, however, remains deterministic in that we cannot study directly probabilistic features of our detection strategy, such as error probabilities, average detection times, etc. This is in contrast with most failure detection techniques where the formulation is based on stochastic models. Yet the types of detection strategies we obtain are similar in many ways to those obtained for stochastic models. For example, our principal on-line detection filter is an extension of the standard Kalman filter, which is the principal tool used for failure detection when stochastic models are used. This shows that there is a close connection between our methodology and previous work based on stochastic modeling. The study of this connection can lead to a better understanding of the probabilistic properties of our detection filter and thus provide a framework in which we can better optimize the compromise between robustness and performance.

Hybrid Stochastic-Deterministic Systems

In many applications the disturbances and model perturbations are best modeled as a combination of stochastic and nonstochastic terms. That is, some of the disturbances are best viewed as random variables and others are best viewed as unknown L^2 functions, as we have frequently done in this book. As noted earlier, a purely stochastic approach lacks robustness and cannot give guaranteed detection as we do in this book. On the other hand, if a disturbance is, say, white noise, then as we have illustrated by examples, if we apply the approach of this book, the result can be conservative in the sense that it will use a larger than necessary auxiliary signal. This can be inefficient or lead to larger perturbations of the state than necessary. Even though we have discussed how a posteriori tuning can be used to adjust the level of performance and robustness if the noise is white (or can be assumed white using standard state augmentation techniques), our approach does not provide a satisfactory paradigm for optimizing the confliciting performance-robustness criteria. One possible way to do so would be the use of the mixed $\mathcal{H}_2/\mathcal{H}_\infty$ criterion introduced by Bernstein and Haddad in the context of control [7]. However, the extension of our result to this case is nontrivial.

Chapter Six

Scilab Programs

6.1 INTRODUCTION

In this chapter some Scilab programs are given to show how the method developed in chapter 3 can be implemented in practice. These are not industrial grade programs. Rather they are given to illustrate the methodology and to solve simple examples. They should, however, be useful for constructing the solution to reasonable sized, well-posed problems. We have tested them on a number of examples. Some large or near-singular problems may require adjusting certain parameters.

6.2 RICCATI-BASED SOLUTION

The Riccati-based solution which was described in section 3.3.3 has been implemented in Scilab. Here we explain how this is done and comment on the code.

This method is based on the computation of the solution to the Riccati equation (3.3.18). This is done using the built-in Scilab function `ode` which solves the ODE defined by the Scilab function `Ricci`. The function `rio` computes the coefficients of the Riccati equation. These coefficients depend on λ and β (via J_β) so they are evaluated frequently. The function `lbcalc` computes λ_β.

To use the functions defined below, the user must have defined the sizes of x and v in the current Scilab environment as `nx` and `nv`, respectively.

```
function lam=lbcalc(beta,P0,P1,J0,J1,A,B,C,D,M,N,prec)
//this function computes lambda for given beta
//using a bisection method
```

```
function [Af,Qf,Rf,Q,R,S]=rio(lam,A,B,C,D,M,N,Jb)
//this function returns the parameters of the Riccati
//equation Af,Qf and Rf
```

```
function pd=Ricci(t,p)
  pd=Af*p+p*Af'-p*Rf*p+Qf;
endfunction
```

The matrices A, B, C, D, M, N used in the problem formulation (3.3.7) are obtained from the two candidate models. Assuming these models are in the form (2.3.21) and using (3.3.5), a Scilab function `datas`, which generates all the matrices needed for the algorithms, can be written as follows.

```
function [A,B,C,D,M,N,F,E0,E1]=..
    datas(A0,B0,C0,D0,G0,H0,M0,N0,A1,B1,C1,D1,G1,H1,M1,N1)
    //Computes augemented system from Models 0 and 1
    [ny,nnu0]=size(N0);[ny,nnu1]=size(N1);
    [nz1,nx1]=size(G1);  [nz0,nx0]=size(G0);
    E0=[zeros(nz0,ny);eye(ny,ny)];E1=[zeros(nz1,ny);eye(ny,ny)];
    M0=[M0,zeros(nx0,nz0)];M1=[M1,zeros(nx1,nz1)];
    N0=[zeros(nz0,nnu0),-eye(nz0,nz0);N0,zeros(ny,nz0)];
    N1=[zeros(nz1,nnu1),-eye(nz1,nz1);N1,zeros(ny,nz1)];
    C0=[G0;C0];C1=[G1;C1];
    D0=[H0;D0];D1=[H1;D1];
    F=kernel([E0;E1]')';F0=F(:,1:nz0+ny);F1=F(:,1+nz0+ny:$);
    B=[B0;B1];A=sysdiag(A0,A1);M=sysdiag(M0,M1);
    C=[F0*C0,F1*C1];D=F0*D0+F1*D1;N=[F0*N0,F1*N1];
endfunction
```

Note that `datas` also returns the matrices F, E_0, and E_1 to be used for the computation of the separating hyperplane.

A Scilab program for solving the suspension problem considered in example 3.7 with time-varying model uncertainty is given below. In order to run these programs the functions described above are supposed to be defined in the file `function.sci`.

```
//scilab script for the suspension example
//using the Riccati-based method

//loading the necessary functions
getf function.sci

T=36; //period of test
//suspension model
th=%pi/6;M=1000;r=.5;m=10;J=1;al=260;k=80;g=10;
P=[M,-M*r*sin(th);-M*r*sin(th),(m+M)*r^2+J];
Q=[al,0;0,0];R=[k,0;0,0];

deltabar=.15;//the uncertainty limit
//Model0 and Model1
A0=[zeros(2,2),eye(2,2);-inv(P)*R,-inv(P)*Q];
B0=[zeros(2,1);inv(P)*[0;1]];
C0=[1,0,0,0;0,0,0,1];
D0=zeros(C0*B0);
M0=[[zeros(2,1);inv(P)*[0;1]],zeros(4,2)];
N0=[[0;0],eye(2,2)]*.1;
G0=[0,0,0,0];H0=deltabar;
A1=A0;B1=B0;
C1=C0;C1(1,1)=0;//C matrices are different
D1=D0;M1=M0;N1=N0;G1=G0;H1=H0;


//computation of sizes
```

```
nz0=size(G0,1);nz1=size(G1,1);
nv=1;nnu0=3;nnu1=3;nx0=4;nx1=4;ny=2;
nzy0=ny+nz0;nzy1=ny+nz1;nx=nx0+nx1;
//defintion of J and P matrices
J0=sysdiag(eye(nnu0,nnu0),-eye(nz0,nz0));
J1=sysdiag(eye(nnu1,nnu1),-eye(nz1,nz1));
P0=eye(nx0,nx0)*.01;P1=eye(nx1,nx1)*.01;

//construction of the augmented system
[A,B,C,D,M,N,F,E0,E1]=..
    datas(A0,B0,C0,D0,G0,H0,M0,N0,A1,B1,C1,D1,G1,H1,M1,N1);

//optimization over beta
n=30; //number of grid points used
lb=[];d=1/n;
a=d/5;b=1-d/5;
bb=linspace(a,b,n)
for beta=bb
  lb=[lb lbcalc(beta)];
end
//plot of the lambda_beta function
xset('window',10)
xbasc();plot2d(bb,lb)
xtitle(' ','beta','lambda')

[lam,i]=max(lb); //computes mximizing beta
if lam==0 then error('No proper auxiliary signal exists');end
beta=bb(i);
//recompute augmented system for optimal beta and lambda
Pb=sysdiag(P0/beta,P1/(1-beta));
Jb=sysdiag(beta*J0,(1-beta)*J1);
lam=lbcalc(beta,.0001)
[Af,Qf,Rf,Q,R,S]=rio(lam);

Ng=5000;
tt=[0:Ng]'*(20+T)/Ng;
PI=ode(Pb,0,tt,Ricci);
//Find a P near the end of the interval
//and compute the kernel of its inverse
mm=size(PI,2)/nx;
disp([T,tt(mm)])
Nn=mm;
tt=linspace(0,tt(mm),Nn); //
//
Pm=inv(PI(:,$-nx+1:$));
eps=.001;
done=%f;
while ~done
  xtm=kernel(Pm,eps,'svd');
  if size(xtm,2)==0 then
```

```
      eps=2*eps;
   elseif size(xtm,2)>=2 then
      eps=eps/2;
   else
      done=%t;
   end
end
// xtm is x_T in (3.3.64)
//back integrate using (3.3.21)
jj=3;//this indicates how far to go--problem depedent
XLAM=ode([xtm;0*xtm],tt($),tt($:-1:$-jj),1d-8,1d-10,bakeq);
LAM=[zeros(nx,nx),eye(nx,nx)]*XLAM;
//continue backward integration using (3.3.65)
lala=ode(LAM(:,$),tt($-jj),tt($-jj-1:-1:1),tpbmodel);

LAM=[LAM,lala];
LAM=LAM(:,$:-1:1);//LAM contains the solution xsi
//Pl contains the solution x obtained from
//x=P*xsi, P is obtained by interpolation of values
//stored in PI
Pl=[];i=1;for t=tt;Pl=[Pl,intrp(t,PI)*LAM(:,i)];i=i+1;end
//optimal auxiliary signal is obtained from (3.3.26)
vv=(-B'*LAM+D'*inv(R)*(C*Pl+S'*LAM))/lam;
nvv=size(vv,2);
tr=[0:nvv-1]'*T/nvv;
alp=sqrt((inv(lam))/(((tr(2)-tr(1))*sum(diag(vv*vv')))));
vv=alp*vv;
vv=vv';
tr=[0:nvv-1]'*T/nvv;
xset('window',30)
xbasc()
plot2d(tr,vv)
xtitle('Auxiliary signal','t','v')

//Computation of the hyperplane parameters
hh=(F*[E0;-E1])'*inv(R)*(C*Pl+S'*LAM);
hh=hh';
tr=[0:nvv-1]'*T/nvv;

NN0=[zeros(nz0,nnu0),-eye(nz0,nz0);N0,zeros(ny,nz0)];
NN1=[zeros(nz1,nnu1),-eye(nz1,nz1);N1,zeros(ny,nz1)];
CC0=[G0;C0];CC1=[G1;C1];
DD0=[H0;D0];DD1=[H1;D1];
Jb=sysdiag(beta*J0,(1-beta)*J1);
NN=sysdiag(NN0,NN1);DD=[DD0;DD1];
Om=[NN,DD]*inv(sysdiag(Jb,-lam*eye(nv,nv)))*[M,B;NN,DD]';
yst=alp*(1/[E0;E1])*(sysdiag(CC0,CC1)*Pl-..
     Om*[-LAM;F'*inv(R)*(C*Pl+S'*LAM)]);
tresh=sum(diag(yst*hh))*(tr(2)-tr(1));
```

```
for i=1:size(hh,2)
  xset('window',i+30)
  xbasc()
  plot2d(tr(1:$),hh(1:$,i))
  xtitle('Hyperplane test','t','h'+string(i))
end
```

The script above assumes that the function `tpbmodel` used for integrating the complete two-point boundary value system (3.3.34), the function `bakeq` used for integrating the backward system (3.3.69), and the function `intrp` used for the computation of $P(t)$ by interpolation, are also included in the file `function.sci`. These functions are given below.

```
function Fd=bakeq(t,F)
//two-point boudary value system (3.3.20)
  Fd=[Af,Qf;Rf,-Af']*F
endfunction
```

```
function dla=tpbmodel(t,la)
//backward system (3.3.65)
  dla=(-Af'+Rf*intrp(t,PI))*la
endfunction
```

```
function PIt=intrp(t,PI)
//returns P(t) by interpolation
kk=max(find(t>tt))
if kk==[] then
  PIt=PI(:,1:nx);return
elseif kk==size(tt,'*') then
    PIt=PI(:,$-nx+1:$);return
end
k=int(kk)
a=(kk)-k;
PIt=(1-a)*PI(:,1+nx*(k-1):nx*k)+a*PI(:,1+nx*(k):nx*(k+1))
endfunction
```

6.3 THE BLOCK DIAGONALIZATION APPROACH

The block diagonalization approach was introduced in section 3.5.1.2. The implementation is, in fact, simpler than that of the Riccati-based solution of the previous section and it is particularly well suited to the case where the cost is in the form (3.5.1a).

The construction of the augmented system is done as in the previous case. It can be done using the `datas` function introduced previously. But here we also have to consider the F, G, U, and W matrices. These matrices define the general cost function; see (3.5.1a) and (3.5.1b).

The function `lbcalc2` is similar to `lbcalc` which we saw in the previous section, but `lbcalc2` uses a different approach to compute λ_β. It also takes F, G, U, and W as additional arguments. The logic of the program is very similar to `lbcalc`, but the test used for the bisection method is based on the result of lemma 3.5.2 and, in particular, the function `tpbvs`. The matrices needed for this test are computed by the Scilab function `rio2` given below. The variable `nf` denotes the size of F and `nz0` and `nz1` denote, respectively, the sizes of z_0 and z_1.

```
function lam=..
      lbcalc2(beta,P0,P1,J0,J1,A,B,C,D,M,N,F,G,U,W,prec)
  printf('Computing lambda for beta=%f\n",beta)
  [lhs,rhs ]=argn()
  if rhs==15 then prec=.01;end
  Lmin=0.00001;Lmax=10000;
  lmin=Lmin;lmax=Lmax;
  [AA,V0,VT]=rio2(lmax,beta,P0,P1,J0,J1,A,B,C,D,M,N,F,G,U,W);
  t=tpbvs(AA,V0,VT,T);
  if t<T then lam=0;printf('does not work'),   return;end
  while lmax-lmin>prec*lmin
    lam=(lmax+lmin)/2; printf('Trying lambda=%f',lam)
    [AA,V0,VT]=rio2(lam,beta,P0,P1,J0,J1,A,B,C,D,M,N,F,G,U,W);
    t=tpbvs(AA,V0,VT,T);
    if t<T   then
      lmin=lam,printf(' --too small, I only got to %f\n",t)
    else
      lmax=lam,printf(' --I can do better, I got to  %f\n",t)
    end
  end
  lam=lmin;
  if lam==Lmin then lam=0;end
  printf('I take lambda=%f\n',lam)
endfunction

function [AA,V0,VT,B,C,D,Gx,S,R]=..
      rio2(lam,beta,P0,P1,J0,J1,A,B,C,D,M,N,F,G,U,W)
  Pb=sysdiag(P0/beta,P1/(1-beta));
  Jb=sysdiag(beta*J0,(1-beta)*J1);
  T=[M,B;N,D]*inv(sysdiag(Jb,-lam*eye(nv,nv)))*[M,B;N,D]';
  Q=T(1:nx,1:nx);S=T(1:nx,nx+1:$);R=T(nx+1:$,nx+1:$);
  No=kernel(N);
  if min(real(spec(No'*Jb*No))) <= 0 ..
      then error('test 1 failed');end
  D=[N,D];C=[C,zeros(ny+nz1+nz0,nf)];
  A=sysdiag(A,F);B=[M,B;zeros(nf,nnu0+nnu1+nz0+nz1),G];
  Gx=sysdiag(Jb,-lam*eye(nv,nv));
  T=[B;D]*inv(Gx)*[B;D]';
  Q=T(1:nx+nf,1:nx+nf);S=T(1:nx+nf,nx+nf+1:$);
  R=T(nx+nf+1:$,nx+nf+1:$);
  AA=[A-S*inv(R)*C,Q-S*inv(R)*S';
      C'*inv(R)*C+sysdiag(zeros(nx,nx),lam*U),..
      -(A-S*inv(R)*C)'];
  V0=[inv(Pb),zeros(nx,nf),-eye(nx,nx),zeros(nx,nf);
```

```
          zeros(nf,nx),eye(nf,nf),zeros(nf,nx+nf);
          zeros(nx+nf,2*nx+2*nf)];
  VT=[zeros(nx+nf,2*nx+2*nf);
          zeros(nx,nx+nf),eye(nx,nx),zeros(nx,nf);
          zeros(nf,nx),-lam*W,zeros(nf,nx),eye(nf,nf)];
endfunction

function T=tpbvs(A,V0,VT,T)
//used for the test in Lemma 3.5.2
//it returns the shortest time interval
//for which the TPBVS becomes singular
  n=size(A,2);
  [Aj ,X,Ab,Af,h]=matbdiag(A);
  V0=V0*X;VT=VT*X;
  II=eye(Ab);JJ=eye(Af);
  nn=size([II(:);JJ(:)],'*');
  [PPh,rd]=ode('root',[II(:);JJ(:)],0,linspace(0,T,20),..
          sys,1,sysr);
  PPh=PPh($-nn+1:$);
  if rd<>[] then
    T=rd(1);
  end
endfunction
```

Note that the function **tpbvs** uses the built-in Scilab ODE solver **ode** with the option **root**, which means with root finder (zero-crossing test). The function being integrated is defined in **sys** corresponding to (3.5.35) and the zero crossing surface is in **sysr** corresponding to (3.5.36). They are given below.

```
function phd=sys(t,Ph)
  phb=matrix(Ph(1:h^2),h,h);
  phf=matrix(Ph(h^2+1:$),n-h,n-h);
  phbd=-Ab*phb
  phfd=Af*phf;
  phd=[phbd(:);phfd(:)];
endfunction

function r=sysr(t,Ph)
  phb=matrix(Ph(1:h^2),h,h);
  phf=matrix(Ph(h^2+1:$),n-h,n-h);
  r=det([sysdiag(eye(Ab),-phf),sysdiag(-phb,eye(Af));V0,VT]);
endfunction
```

The function **matbdiag** given below block diagonalizes H as indicated in equation (3.5.32). There are various ways to do the diagonalization. The method used in the following program is to triangularize the matrix using the Schur method and then block diagonalize using a Sylvester equation. See below.

```
function [A,X,A1,A3,dim]=matbdiag(AA)
  [U,dim,T ]=schur(AA,extern1)
```

```
  A1=T(1:dim,1:dim);A2=T(1:dim,dim+1:$);A3=T(dim+1:$,dim+1:$);
  Y = sylv(A1,-A3,A2,'c');
  X=U*[eye(A1),-Y;zeros(A2'),eye(A3)];
  bs=[dim,size(A,1)-dim];
  A=sysdiag(A1,A3);
endfunction

function s=extern1(Ev)
  s=%f;
  if real(Ev)>.00001 then s=%t,end
endfunction
```

6.4 GETTING SCILAB AND THE PROGRAMS

Scilab may be downloaded form the French research center INRIA (Institut National de Recherche en Informatique et en Automatique). The Scilab website is at `http://www.scilab.org/`.

At `http://www4.ncsu.edu/~slc/www/BOOKS/AuxSig.html` there are a number of items available for download. These include the listings of the programs given in this chapter, some additional programs, and the Scilab code for different specific examples. Once the new version of Scilab with Scicos is released, there will also be Scicos diagrams showing how to set up some of the applications mentioned in this book. Scicos is a graphical interface to Scilab that allows for modeling with block diagrams. Its relationship to Scilab is similar to that of Simulink to MATLAB. (MATLAB and Simulink are copyrighted products of The MathWorks, Inc.) Information on Scicos can be found at `http://www.scicos.org`.

Suggestions from readers for additional information to include on the website are welcome as are suggestions for links to other information. The authors may be reached at `slc@math.ncsu.edu` and `ramine.nikoukhah@inria.fr`.

Appendix A

List of Symbols

The following is a list of symbols and notation used in this book. For general concepts the definition may be given. In cases where the notation is specific to this book the page on which it is first defined is also given. Notation that appears only once in a given section may not be repeated here.

Symbol	Page	Meaning
$\forall$		For every
$\mathbb{C}$		Complex numbers
x		State
y		Output
ν		Noise
x_k		Discrete state
y_k		Discrete output
δX		Uncertainty in matrix X
ν_k		Discrete noise
$\mathcal{A}_i(v)$		Output set for model i given signal v
$\mathcal{L}_A(f)$		Solution of $\dot{x} = Ax + f, x(0) = 0$
$E^\perp$		Largest full row rank matrix such that $E^\perp E = 0$
$E_\perp$		Largest full column rank matrix such that $EE_\perp = 0$
$E^T_\perp$	22	Same as $(E^T)_\perp$
$E^\dagger$		Moore-Penrose pseudoinverse of E
E^r		Right inverse of E $(EE^r = I)$
E^l		Left inverse of E $(E^l E = I)$
E^T		Transpose of E
$\dot{E}$		Derivative with respect to scalar t of $E(t)$
E_x		Derivative with respect to scalar or vector x of vector function E
$\bar{\sigma}(A)$		Largest singular value of the matrix A
$\mathrm{diag}(a_1, \dots, a_n)$		Block diagonal matrix whith a_i on diagonal.
$S_i(\nu, x_i(0))$	123	Disturbance measure for model i
$S_i(\nu)$	123	Disturbance measure for model i with no weight on $x_i(0)$
$S_i(\nu, s)$	76	Disturbance measure with model uncertainty
γ^*	63	Separability index
$\ker(A)$		Nullspace of matrix A

Symbol	Page	Meaning
$\|v\|$		Absolute value of a number or norm of a vector
$\|v\|$		L^2 norm of a vector function v: $\|v\|^2 = \int_0^T \|v(s)\|^2 ds$
$\|v\|_{[\tau,T]}$		L^2 norm of a vector function v on interval $[\tau, T]$
$\|v\|_\tau$		Same as $\|v\|_{[0,\tau]}$
$\|H\|$		sup norm of a matrix valued function $H(t)$
$\langle f, g \rangle$		Inner product in L^2 of functions $f(t), g(t)$
$\rho(A)$		Spectral radius of matrix A
$\text{Im}(A)$		Image (range) of matrix A
$\sigma(v)$	61	inf max of noise bounds, static case
$\sigma(v,s)$	77	inf max of noise bounds, continuous model uncertainty
$\sigma(v,k)$	99	inf max of noise bounds, discrete model uncertainty
$\phi_\beta(v)$	62	Value of auxiliary problem given β, v
$\phi_\beta(v,s)$	77	Auxiliary problem value with continuous model uncertainty
$\phi_\beta(v,k)$	100	Auxiliary problem value with discrete model uncertainty
$\phi_\beta^{ij}(v)$	139	
$\mathcal{B}$	78	
$\mathcal{B}(v,s)$	79	
J_i	60	Signature matrix in the noise bound for model i
J_β	63	
J	22	Signature matrix in the noise bound
$\phi_{i,j}(y)$	144	
$H(u,y), \mu$	14	Residual
Δ	20	Matrix perturbation parameter
$\mathcal{E}$		Expected value in chapter 3. Matrix in chapter 4
$\gamma(u,y)$	16	
$\gamma_s(u,y)$	29	
v		Auxiliary signal
$\lambda_{\beta,s}$	80	
det		Determinant
P_β	77	
$\mathcal{V}$	63	Set of proper v
V_β	64	
v^*	69	Optimal proper v
λ^*	64	λ for v^*
β^*	69	β for v^*
$\phi_\delta(y)$	131	
ϕ_ϵ	133	
a_ϵ	133	Approximate normal in hyperplane
$\mathcal{L}_i$	124	Same as $\mathcal{L}_{A_i}$

Symbol	Page	Meaning
$\hat{x}$		Estimate of x
$J(a,c)$	44	
$J(s,a,c)$	47	
$R_{\lambda\beta}$	80	
$Q_{\lambda\beta}$	80	
$S_{\lambda\beta}$	80	
$\oplus$	113	Direct sum of subspaces
$G_i(x)$	112	Quadratic forms
$\Gamma_{\beta,\lambda}$	94	
$\delta(v)$	123	General norm of v
V_i	95	
V_f	95	
$O(h)$		$O(h)/h$ is bounded for small nonzero h
$A \otimes B$		Kronecker product of matrices A, B
L^2		Functions f for which $\|f\| < \infty$
$Q \geq 0$		Q is a positive semidefinite matrix
$Q > 0$		Q is a positive definite matrix
$P \leq Q$		$Q - P \geq 0$ for matrices P, Q

Bibliography

[1] M. Basseville. Detecting changes in signals and systems–A survey. *Automatica*, 24(3):309–326, 1988.

[2] M. Basseville, M. Abdelghani, and A. Benveniste. Subspace-based fault detection algorithms for vibration monitoring. *Automatica*, 36:101–109, 1999.

[3] M. Basseville and A. Benveniste, editors. *Detection of Abrupt Changes in Signals and Dynamical Systems*, volume 77 of *Lecture Notes in Control and Information Sciences*. Springer, 1985.

[4] M. Basseville and I. V. Nikiforov. *Detection of Abrupt Changes: Theory and Application*. Information and System Science Series. Prentice-Hall, Englewood Cliffs, N.J., 1993.

[5] R. V. Beard. *Failure Accommodation in Linear Systems through Self-reorganization*. PhD thesis, Massachusetts Insitute of Technology, Cambridge, MA, 1971.

[6] R. E. Bellman. *Dynamic Programming*. Princeton University Press, Princeton, NJ, 1957.

[7] D. S. Bernstein and W. M. Haddad. LQG control with an H_∞ performance bound: A Riccati equation approach. *IEEE Trans. Autom. Control*, 34:293–305, 1989.

[8] D. P. Bertsekas and I. B. Rhodes. Recursive state estimation for a set-membership description of uncertainty. *IEEE Trans. Autom. Control*, AC-16(2):117–128, 1971.

[9] J. T. Betts. *Practical Methods for Nonlinear Control Using Nonlinear Programming*, volume 3 of *Advances in Design and Control*. SIAM, Philadelphia, PA, 2000.

[10] J. T. Betts, N. Biehn, S. L. Campbell, and W. P. Huffman. Compensating for order variation in mesh refinement for direct transcription methods. *J. Comput. Appl. Math.*, 125:147–158, 2000.

[11] J. T. Betts, N. Biehn, S. L. Campbell, and W. P. Huffman. Compensating for order variation in mesh refinement for direct transcription methods II: Computational experience. *J. Comput. Appl. Math.*, 143:237–261, 2002.

[12] J. T. Betts, M. J. Carter, and W. P. Huffman. Software for nonlinear optimization. Technical report, Mathematics and Engineering Analysis, Boeing Information and Support Services, The Boeing Company, June 1997.

[13] J. T. Betts and P. D. Frank. A sparse nonlinear optimization algorithm. *J. Optim. Theory Applic.*, 82:519–541, 1994.

[14] J. T. Betts and W. P. Huffman. Sparse optimal control software. Technical report, Mathematics and Engineering Analysis, Boeing Information and Support Services, The Boeing Company, Seattle, June 1997.

[15] K. E. Brenan, S. L. Campbell, and L. R. Petzold. *The Numerical Solution of Initial Value Problems in Differential-Algebraic Equations*. SIAM, Philadelphia, PA, 1996.

[16] C. Bunks, J. P. Chancelier, F. Delebecque, C. Gomez, M. Goursat, R. Nikoukhah, and S. Steer, editors. *Engineering and Scientific Computing with Scilab*. Birkhauser, Basel, 1999.

[17] S. L. Campbell, K. J. Drake, and R. Nikoukhah. Auxiliary signal design for multi-model identification in systems with multiple delays. In *Proceedings of the IEEE Mediterranean Conference on Control and Automation*, Lisbon, 2002. (CD ROM).

[18] S. L. Campbell, K. J. Drake, and R. Nikoukhah. Early decision making when using proper auxiliary signals. In *Proceedings of the IEEE Conference on Decision and Control*, pages 1832–1837, Las Vegas, NV, 2002.

[19] S. L. Campbell, K. J. Drake, and R. Nikoukhah. Analysis of spline based auxiliary signal design for failure detection in delay systems. In *Proceedings of the IEEE Conference on Systems, Man, and Cybernetics*, Washington, DC, 2003. (CD ROM).

[20] S. L. Campbell, K. J. Drake, R. Nikoukhah, and F. Delebecque. Rapid multi-model identification in systems with delay. In *Proceedings of the 3d IFAC Workshop on Time Delay Systems*, Sante Fe, NM, 2001. (CD ROM).

[21] S. L. Campbell, K. Horton, R. Nikoukhah, and F. Delebecque. Rapid model selection and the separability index. In *Proceedings of the IFAC Symposium on Fault Detection Supervision and Safety for Technical Processes (SAFEPROCESS 2000)*, pages 1187–1192, Budapest, Hungary, 2000.

[22] S. L. Campbell, K. Horton, R. Nikoukhah, and F. Delebecque. Auxiliary signal design for rapid multimodel identification using optimization. *Automatica*, 38:1313–1325, 2002.

[23] S. L. Campbell and C. D. Meyer, Jr. *Generalized Inverses of Linear Transformations*. Dover, New York, 1991.

[24] S. L. Campbell and R. Nikoukhah. The design of auxiliary signals for robust active failure detection in uncertain systems. In *Proceedings of the Mathematical Theory of Networks and Systems*, Notre Dame, IN, 2002. (CD ROM).

[25] J. P. Chancelier, F. Delebecque, C. Gomez, M. Goursat, R. Nikoukhah, and S. Steer. *Introduction à Scilab*. Springer, 2002.

[26] J. Chen and R. J. Patton. *Robust Model–Based Fault Detection Diagnosis for Dynamic Systems*. Kluwer, Dordrecht, 1999.

[27] L. H. Chiang, E. L. Russell, and R. D. Braatz. *Fault Detection and Diagnosis in Industrial Systems*. Springer, New York, 2001.

[28] W. H. Chung and J. L. Speyer. A game theoretic fault detection filter. *IEEE Trans. Automat. Contr.*, AC-43(2), 1998.

[29] L. Dai. *Singular Control Systems*. Lecture Notes in Control and Information Sciences. Springer, Berlin, 1989.

[30] X. Ding and P. M. Frank. Fault detection via factorization approach. *Syst. Control Lett.*, 14:431–436, 1990.

[31] R. K. Douglas and J. L. Speyer. An H_∞ bounded fault detection filter. In *Proceedings of the American Control Conference*, pages 86–90, Seattle, 1995.

[32] K. Drake. *Analysis of Numerical Methods for Fault Diagnosis and Model Identification in Linear Systems with delays*. PhD thesis, North Carolina State University, Raleigh, NC, 2003.

[33] A. Edelmayer, J. Boker, and L. Keviczky. H_∞ detection filter design for linear systems: comparison of two approaches. In *Proceedings of the IFAC World Congress*, pages 37–42, San Francisco, CA, 1996.

[34] P. M. Frank. Fault diagnosis in dynamic systems using analytic and knowledge-based redundancy–A survey and some new results. *Automatica*, 26(3):259–474, 1990.

[35] P. M. Frank. On-line fault detection in uncertain nonlinear systems using diagnostic observers: A survey. *Int. J. Syst. Sci.*, 25:2129–2154, 1994.

[36] P. M. Frank. Analytical and qualitative model based fault diagnosis – A survey and some new reults. *Eur. J. Control*, 2:6–28, 1996.

[37] E. A. Garcia and P. M. Frank. Deterministic nonlinear observer-based approaches to diagnosis: A survey. *Control Eng. Pract.*, 5:663–670, 1997.

[38] J. J. Gertler. Survey of model-based failure detection and isolation in complex plants. *IEEE Control Syst. Mag.*, 12:3–11, 1988.

[39] J. J. Gertler. *Fault Detection and Diagnosis in Engineering Systems*. Marcel Dekker, New York, 1998.

[40] L. El Ghaoui and G. Calafiore. Confidence ellipsoids for uncertain linear equations with structure. In *Proceedings of the IEEE Conference on Decision and Control*, pages 1922–1927, Phoenix, AZ, 1999.

[41] L. El Ghaoui and G. Calafiore. Robust filtering for discrete-time systems with structured uncertainty. *IEEE Trans. Autom. Control*, AC-46(7):1084–1089, 2001.

[42] L. El Ghaoui and H. Lebret. Robust solutions to least-squares problems with uncertain data. *SIAM J. Matrix Anal. Appl.*, 18:1035–1064, 1997.

[43] K. Horton. *Fault Diagnosis and Model Identification in Linear Descriptor Systems*. PhD thesis, North Carolina State University, Raleigh, 2001.

[44] R. Isermann. Process fault detection based on modeling and estimation methods – A survey. *Automatica*, 20(4):387–404, 1984.

[45] K. Ito and F. Kappel. A uniformly differentiable approximation scheme for delay systems using splines. *Appl. Math. Optim.*, 23:217–262, 1991.

[46] H. L. Jones. *Failure Detection in Linear Systems*. PhD thesis, Massachusetts Institute of Technology, Cambridge, MA, 1973.

[47] T. Kailath. *Linear Systems*. Prentice-Hall, Englewood Cliffs, NJ, 1980.

[48] F. Kerestecioğlu. *Change Detection and Input Design in Dynamical Systems*. Research Studies Press, Baldock, Hertfordshire, 1993.

[49] F. Kerestecioğlu and M. B. Zarrop. Optimal input design for change detection in dynamical systems. In *Proceedings of the European Control Conference*, pages 321–326, Grenoble, 1991.

[50] F. Kerestecioğlu and M. B. Zarrop. Input design for detection of abrupt changes in dynamical systems. *Int. J. Control*, 59:1063–1084, 1994.

[51] M. Kinnaert, Y. Peng, and H. Hammouri. The fundamental problem of residual generation for bilinear systems up to output injection. In *Proceedings of the European Control Conference*, pages 3777–3782, Roma, 1995.

[52] A. Kurzhanski and I. Valyi. *Ellipsoidal Calculus for Estimation and Control*. Birkhauser, 1997.

[53] X-C. Lou, A. S. Willsky, and G. C. Verghese. Optimally robust redundancy relations for failure detection in uncertain systems. *Automatica*, 22(3):333–344, 1986.

[54] R. S. Mangoubi. *Robust Estimation and Failure Detection: A Concise Treatment*. Advances in Industrial Control. Springer, Berlin, 1998.

[55] A. Manitius and H. Tran. Numerical simulation of a non-linear feedback controller for a wind tunnel model involving a time delay. *Opt. Control Appl. Methods*, 7:19–39, 1986.

[56] M. A. Massoumnia. *A Geometric Approach to Failure Detection and Identi-fication in Linear Systems*. PhD thesis, Massachusetts Insitute of Technology, Cambridge, MA, 1986.

[57] M. A. Massoumnia. A geometric approach to the synthesis of failure detection filters. *IEEE Trans. Autom. Control*, AC-31(9):839–846, 1986.

[58] M. A. Massoumnia, G. C. Verghese, and A. S. Willsky. Failure detection and identification. *IEEE Trans. Autom. Control*, AC-34(3):316–321, 1989.

[59] R. K. Mehra and J. Peschon. An innovations approach to fault detection and diagnosis in dynamic systems. *Automatica*, 7:637–640, 1971.

[60] H. H. Niemann and J. Stoustrup. Integration of control and fault detection: Nominal and robust design. In *Proceedings of the Third IFAC Symposium on Fault Detection, Supervision and Safety for Technical Processes (SAFEPRO-CESS 1997)*, pages 341–346, Hull, England, 1997.

[61] R. Nikoukhah. Innovations generation in the presence of unknown inputs, ap-plication to robust failure detection. *Automatica*, 30(12):1851–1867, 1994.

[62] R. Nikoukhah. Guaranteed active failure detection and isolation for linear dy-namical systems. *Automatica*, 34(11):1345–1358, 1998.

[63] R. Nikoukhah and S. L. Campbell. Active failure detection: Auxiliary sig-nal design and on-line detection. In *Proceedings of the IEEE Mediterranean Conference on Control and Automation*, Lisbon, 2002. (CD ROM).

[64] R. Nikoukhah and S. L. Campbell. Auxiliary signal design for failure detection in uncertain sampled-data systems. In *Proceedings of the European Control Conference*, Cambridge, UK, 2003.

[65] R. Nikoukhah and S. L. Campbell. Auxiliary signal design for robust active fail-ure detection: The general cost case. In *Proceedings of the 5th IFAC Symposium on Fault Detection, Supervision, and Safety of Technical Processes (SAFEPRO-CESS 2003)*, pages 259–264, Washington, DC, 2003.

[66] R. Nikoukhah, S. L. Campbell, and F. Delebecque. Detection signal design for failure detection: a robust approach. *Int. J Adap. Control Signal Process.*, 14:701–724, 2000.

[67] R. Nikoukhah, S. L. Campbell, K. Horton, and F. Delebecque. Auxiliary sig-nal design for robust multi-model identification. *IEEE Trans. Autom. Control*, 47:158–163, 2002.

[68] R. J. Patton and J. Chen. A survey of robustness in quantitative model-based fault diagnosis. *J. Appl. Math. Comput. Sci.*, 3:15–32, 1993.

[69] R. J. Patton, P. M. Frank, and R. N. Clark. *Fault Diagnosis in Dynamic Systems–Theory and Application*. Prentice-Hall, Englewood Cliffs, NJ, 1989.

[70] R. J. Patton, P. M. Frank, and R. N. Clark, editors. *Issues of Fault Diagnosis for Dynamic Systems*. Springer, Berlin, 2000.

[71] I. R. Petersen and D. C. McFarlane. A methodology for process fault detection. In *Proceedings of the IEEE Conference on Decision and Control*, pages 4984–4989, Phoenix, AZ, 1999.

[72] I. R. Petersen and A. V. Savkin. *Robust Kalman Filtering for Signals and Systems with Large Uncertainties*. Advances in Design and Control. Birkhauscr, 1999.

[73] I. R. Petersen, V. A. Ugrinovskii, and A. V. Savkin. *Robust Control Design Using H^∞ Methods*. Communications and Control Engineering. Springer, Berlin, 2000.

[74] W. T. Reid. *Sturmain Theory for Ordinary Differential Equations*. Springer, New York, 1990.

[75] A. V. Savkin and I. R. Petersen. New approach to model validation and fault diagnosis. *J. Opt. Theory Appl.*, 94(1):241–250, 1997.

[76] A. V. Savkin and I. R. Petersen. Robust state estimation and model validation for discrete-time uncertain systems with a deterministic description of noise uncertainty. *Automatica*, 34(2):271–274, 1998.

[77] A. H. Sayed. A framework for state-space estimation with uncertain models. *IEEE Trans. Autom. Control*, AC-46(7):998–1013, 2001.

[78] F. C. Schweppe. Recursive state estimation: Unknown but bounded errors and system inputs. *IEEE Trans. Autom. Control*, AC-13(1):1017–1018, 1968.

[79] F. C. Schweppe. *Uncertain Dynamic Systems*. Prentice Hall, Englewood Cliffs, NJ, 1973.

[80] M. Tyler and M. Morari. Optimal and robust design of integrated control and diagnostic modules. In *Proceedings of the American Control Conference*, volume 2, pages 2060–2064, Baltimore, MD, 1994.

[81] E. Wahnon, A. Benveniste, L. El Ghaoui, and R. Nikoukhah. An optimum robust approach to statistical failure detection and identification. In *Proceedings of the IEEE Conference on Decision and Control*, pages 650–655, Brighton, 1991.

[82] J. C. Willems, A. Kitapçi, and L. M. Silverman. Singular optimal control: A geometric approach. *SIAM J. Control Optim.*, 24(2):323–337, 1986.

[83] A. S. Willsky. A survey of design methods for failure detection in dynamics system. *Automatica*, 12(6):601–611, 1976.

[84] A. S. Willsky and H. L. Jones. A generalized likelihood ratio approach to the detection and estimation of jumps in linear systems. *IEEE Trans. Autom. Control*, AC-21(1):108–112, 1976.

[85] H. Yang and M. Saif. State observation, failure detection and isolation (FDI) in bilinear systems. In *Proceedings of the IEEE Conference on Decision and Control*, pages 1391–2396, New Orleans, LA, 1995.

[86] J. S. Yang. Mixed H^2 compensator design for an aircraft control problem. In *Proceedings of the IEEE Conference on Decision and Control*, pages 1964–1969, Phoenix, AZ, 1999.

[87] D. Yu and D. N. Shields. Fault diagnosis in bilinear systems- a survey. In *Proceedings of the Europen Control Conference*, pages 360–365, Roma, 1995.

[88] D. Yu and D. N. Shields. A bilinear fault detection observer. *Automatica*, 32:1597–1607, 1996.

[89] D. Yu and D. N. Shields. A bilinear fault detection filter. *Int. J. Control*, 68(3):417–430, 1997.

[90] X. J. Zhang. *Auxiliary Signal Design in Fault Detection and Diagnosis*. Springer, Heidelberg, 1989.

Index

9 780691 099873